ELSEVIER'S INTEGRATED REVIEW
IMMUNOLOGY AND MICROBIOLOGY

ELSEVIER'S INTEGRATED REVIEW
IMMUNOLOGY AND MICROBIOLOGY
SECOND EDITION

Jeffrey K. Actor, PhD
Professor
Department of Pathology and Laboratory Medicine
University of Texas-Houston Medical School
Houston, Texas

ELSEVIER
SAUNDERS

1600 John F. Kennedy Blvd. Ste 1800
Philadelphia, PA 19103-2899

ELSEVIER'S INTEGRATED REVIEW IMMUNOLOGY
AND MICROBIOLOGY, SECOND EDITION

ISBN: 978-0-323-07447-6

Copyright © 2012 by Saunders, an imprint of Elsevier Inc.
Copyright © 2007 by Mosby, Inc., an affiliate of Elsevier Inc.

No part of this publication may be reproduced or transmitted in any form or by any means, electronic or mechanical, including photocopying, recording, or any information storage and retrieval system, without permission in writing from the publisher. Details on how to seek permission, further information about the Publisher's permissions policies and our arrangements with organizations such as the Copyright Clearance Center and the Copyright Licensing Agency, can be found at our website: www.elsevier.com/permissions.

This book and the individual contributions contained in it are protected under copyright by the Publisher (other than as may be noted herein).

Notices

Knowledge and best practice in this field are constantly changing. As new research and experience broaden our understanding, changes in research methods, professional practices, or medical treatment may become necessary.

Practitioners and researchers must always rely on their own experience and knowledge in evaluating and using any information, methods, compounds, or experiments described herein. In using such information or methods they should be mindful of their own safety and the safety of others, including parties for whom they have a professional responsibility.

With respect to any drug or pharmaceutical products identified, readers are advised to check the most current information provided (i) on procedures featured or (ii) by the manufacturer of each product to be administered, to verify the recommended dose or formula, the method and duration of administration, and contraindications. It is the responsibility of practitioners, relying on their own experience and knowledge of their patients, to make diagnoses, to determine dosages and the best treatment for each individual patient, and to take all appropriate safety precautions.

To the fullest extent of the law, neither the Publisher nor the authors, contributors, or editors, assume any liability for any injury and/or damage to persons or property as a matter of products liability, negligence or otherwise, or from any use or operation of any methods, products, instructions, or ideas contained in the material herein.

Library of Congress Cataloging-in-Publication Data

Actor, Jeffrey K.
  Elsevier's integrated review immunology and microbiology / Jeffrey K. Actor. – 2nd ed.
    p. ; cm.
  Integrated review immunology and microbiology
  Rev. ed. of: Elsevier's integrated immunology and microbiology / Jeffrey K. Actor. c2007.
  Includes index.
  ISBN 978-0-323-07447-6 (pbk. : alk. paper)
  I. Actor, Jeffrey K. Elsevier's integrated immunology and microbiology. II. Title. III. Title: Integrated review immunology and microbiology.
  [DNLM: 1. Immune System Phenomena. 2. Microbiological Phenomena. QW 540]

  616.07'9–dc23   2011018052

Acquisitions Editor: Madelene Hyde
Developmental Editor: Andrew Hall
Publishing Services Manager: Patricia Tannian
Team Manager: Hemamalini Rajendrababu
Project Manager: Antony Prince
Designer: Steven Stave

Printed in China

Last digit is the print number: 9 8 7 6 5 4 3 2 1

Working together to grow
libraries in developing countries

www.elsevier.com | www.bookaid.org | www.sabre.org

ELSEVIER   BOOK AID International   Sabre Foundation

To my father, Paul Actor, PhD, who instilled in me a sense of excitement about the wonders of science and the curiosity to ask questions about biological systems

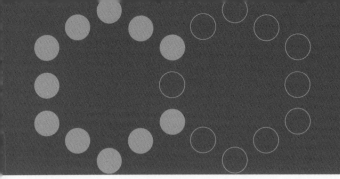

Preface

Immunology represents a rapidly changing field with new theories actively evolving as molecular techniques broaden our scientific perspective on interactions between pathogens and the human host. The immune cells and organs of the body comprise the primary defense system against invasion by microorganisms. A functional immune system confers a state of health through effective immune surveillance and elimination of infectious agents. The study of immunologic and hematologic principles, as applied toward understanding host protection against pathogenic assault, integrates well with microbiology and the study of basic concepts underlying the nature of foreign pathogens. The goal of the first half of the book is to present immune system components, both innate and adaptive, in a concise manner to elucidate their intertwined relationships that culminate in effective host protection and health. The remaining chapters present the world of microbiology, with a concise overview of clinically relevant bacteria, viruses, fungi, and parasites, to allow an understanding of infectious organisms as the causative agents underlying human disease.

This book is aimed at students of human health and those in the medical profession; it is written to simplify concepts and encourage inquisitive individuals to explore further medically relevant topics. Indeed, the purpose of the integrated text series is to encourage cross-disciplinary thought across multiple sciences. Integration boxes promote cross-discipline thinking and allow the reader to build bridges between related ideas in other medical fields. The clinical vignettes (Case Studies) and associated questions at the end of the book are organized to provide perspectives into molecular aspects underlying clinical disease manifestation. These scenarios are aimed to assist in understanding consequences of ineffective, inappropriate, overactive, or nonregulated responses and their relationship to immunologic disorders and deficiencies as well as to responses occurring during infection. The associated USMLE format questions available at www.StudentConsult.com will also test knowledge in a clinical context, with succinct explanations to allow increased application of immunologic and microbiologic concepts to medically related disease states.

Overall, the text attempts to present information in a clinically relevant and focused manner that outlines concepts for further exploration, creating a base of knowledge for those with a desire to understand how the healthy individual combats disease.

Jeffrey K. Actor, PhD

Editorial Review Board

Chief Series Advisor
J. Hurley Myers, PhD
Professor Emeritus of Physiology and Medicine
Southern Illinois University School of Medicine;
President and CEO
DxR Development Group, Inc.
Carbondale, Illinois

Anatomy and Embryology
Thomas R. Gest, PhD
University of Michigan Medical School
Division of Anatomical Sciences
Office of Medical Education
Ann Arbor, Michigan

Biochemistry
John W. Baynes, MS, PhD
Graduate Science Research Center
University of South Carolina
Columbia, South Carolina

Marek Dominiczak, MD, PhD, FRCPath, FRCP(Glas)
Clinical Biochemistry Service
NHS Greater Glasgow and Clyde
Gartnavel General Hospital
Glasgow, United Kingdom

Clinical Medicine
Ted O'Connell, MD
Clinical Instructor
David Geffen School of Medicine
UCLA;
Program Director
Woodland Hills Family Medicine Residency Program
Woodland Hills, California

Genetics
Neil E. Lamb, PhD
Director of Educational Outreach
Hudson Alpha Institute for Biotechnology
Huntsville, Alabama;
Adjunct Professor
Department of Human Genetics
Emory University
Atlanta, Georgia

Histology
Leslie P. Gartner, PhD
Professor of Anatomy
Department of Biomedical Sciences
Baltimore College of Dental Surgery
Dental School
University of Maryland at Baltimore
Baltimore, Maryland

James L. Hiatt, PhD
Professor Emeritus
Department of Biomedical Sciences
Baltimore College of Dental Surgery
Dental School
University of Maryland at Baltimore
Baltimore, Maryland

Immunology
Darren G. Woodside, PhD
Principal Scientist
Drug Discovery
Encysive Pharmaceuticals, Inc.
Houston, Texas

Microbiology
Richard C. Hunt, MA, PhD
Professor of Pathology, Microbiology, and Immunology
Director of the Biomedical Sciences Graduate Program
Department of Pathology and Microbiology
University of South Carolina School of Medicine
Columbia, South Carolina

Neuroscience
Cristian Stefan, MD
Associate Professor
Department of Cell Biology
University of Massachusetts Medical School
Worcester, Massachusetts

Pathology
Peter G. Anderson, DVM, PhD
Professor and Director of Pathology Undergraduate
Education, Department of Pathology
University of Alabama at Birmingham
Birmingham, Alabama

Editorial Review Board

Pharmacology
Michael M. White, PhD
Professor Department of Pharmacology and Physiology
Drexel University College of Medicine
Philadelphia, Pennsylvania

Physiology
Joel Michael, PhD
Department of Molecular Biophysics and Physiology
Rush Medical College
Chicago, Illinois

Acknowledgments

I would like to thank Robert L. Hunter Jr, MD, PhD; Steven J. Norris, PhD; and Gailen D. Marshall, MD, PhD, who supported and encouraged me as I strove to reach my academic and research goals. Special thanks go to Alexandra Stibbe at Elsevier for her vision and insights that made the Integrated Series possible, to Kate Dimock for her excellent leadership in the project, and to Andrew C. Hall for his attention to detail, his humor, and his positive guidance. Finally, I could not have completed this endeavor without the support of my wife Lori and her continued faith in my abilities and encouragement to follow my dreams.

Contents

SECTION I IMMUNOLOGY

1 Introduction to Immunity and Immune Systems — 3
2 Cells and Organs of the Immune System — 7
3 Humoral Immunity: Antibody Recognition of Antigen — 17
4 T-Cell Immunity — 25
5 Role of Major Histocompatibility Complex in the Immune Response — 33
6 Innate Immunity — 43
7 Adaptive Immune Response and Hypersensitivity — 53
8 Immunomodulation — 61
9 Immunoassays — 71
10 Infection and Immunity — 81

SECTION II MICROBIOLOGY

11 Basic Bacteriology — 93
12 Clinical Bacteriology — 105
13 Basic Virology — 121
14 Clinical Virology — 129
15 Mycology — 139
16 Parasitology — 147

Case Studies — 157

Case Study Answers — 161

Index — 167

Series Preface

How to Use This Book

The idea for Elsevier's Integrated Series came about at a seminar on the USMLE Step 1 Exam at an American Medical Student Association (AMSA) meeting. We noticed that the discussion between faculty and students focused on how the exams were becoming increasingly integrated—with case scenarios and questions often combining two or three science disciplines. The students were clearly concerned about how they could best integrate their basic science knowledge.

One faculty member gave some interesting advice: "read through your textbook in, say, biochemistry, and every time you come across a section that mentions a concept or piece of information relating to another basic science—for example, immunology—highlight that section in the book. Then go to your immunology textbook and look up this information, and make sure you have a good understanding of it. When you have, go back to your biochemistry textbook and carry on reading."

This was a great suggestion—if only students had the time, and all of the books necessary at hand, to do it! At Elsevier we thought long and hard about a way of simplifying this process, and eventually the idea for Elsevier's Integrated Series was born.

The series centers on the concept of the integration box. These boxes occur throughout the text whenever a link to another basic science is relevant. They're easy to spot in the—with their color-coded headings and logos. Each box contains a title for the integration topic and then a brief summary of the topic. The information is complete in itself—you probably won't have to go to any other sources—and you have the basic knowledge to use as a foundation if you want to expand your knowledge of the topic.

You can use this book in two ways. First, as a review book . . .

When you are using the book for review, the integration boxes will jog your memory on topics you have already covered. You'll be able to reassure yourself that you can identify the link, and you can quickly compare your knowledge of the topic with the summary in the box. The integration boxes might highlight gaps in your knowledge, and then you can use them to determine what topics you need to cover in more detail.

Second, the book can be used as a short text to have at hand while you are taking your course . . .

You may come across an integration box that deals with a topic you haven't covered yet, and this will ensure that you're one step ahead in identifying the links to other subjects (especially useful if you're working on a PBL exercise). On a simpler level, the links in the boxes to other sciences and to clinical medicine will help you see clearly the relevance of the basic science topic you are studying. You may already to confident in the subject matter of many of the integration boxes, so they will serve as helpful reminders.

At the back of the book we have included case study questions relating to each chapter so that you can test yourself as you work your way through the book

Online Version

An online version of the book is available on our Student Consult site. Use of this site is free to anyone who has bought the printed book. Please see the inside front cover for full details on Student Consult and how to access the electronic version of this book.

In addition to containing USMLE test questions, fully searchable text, and an image bank, the Student Consult site offers additional integration links, both to the other books in Elsevier's Integrated Series and to other key Elsevier textbooks.

Books in Elsevier's Integrated Series

The nine books in the series cover all of the basic sciences. The more books you buy in the series, the more links that are made accessible across the series, both in print and online.

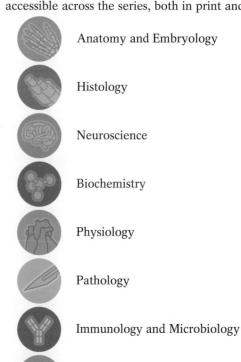

Anatomy and Embryology

Histology

Neuroscience

Biochemistry

Physiology

Pathology

Immunology and Microbiology

Pharmacology

Genetics

SECTION I
Immunology

Introduction to Immunity and Immune Systems 1

CONTENTS
CHIEF FUNCTION OF IMMUNITY
INNATE IMMUNE SYSTEM
ADAPTIVE IMMUNE SYSTEM
SPECIFICITY OF ADAPTIVE RESPONSE BY LYMPHOCYTE RECEPTORS
TIGHT REGULATION OF THE IMMUNE SYSTEM AND ASSOCIATED RESPONSES

CHIEF FUNCTION OF IMMUNITY

The immune cells and organs of the body make up the primary defense system against invasion by microorganisms and foreign pathogens. A functional immune system confers a state of health through effective elimination of infectious agents (bacteria, viruses, fungi, and parasites) and through control of malignancies by protective immune surveillance. In essence, the process is based in functional discernment between self and nonself, a process that begins in utero and continues through adult life.

Immune responses are designed to interact with the environment to protect the host against pathogenic invaders. The goal of the chapters in Section I is to provide an appreciation of the components of the human immune response that work together to protect the host. In addition, a working clinical understanding of the concept of immune-based diseases resulting from either immune system component deficiencies or excess activity will be presented.

INNATE IMMUNE SYSTEM

The immune system consists of two overlapping compartments representing interactions between innate and adaptive components and associated responses. The innate immune mechanisms provide the first line of defense against infectious disease (Table 1-1). Innate immune components are present from birth and consist of nonspecific components available before the onset of infection. Innate immune recognition uses preformed effector molecules to recognize broad structural motifs that are highly conserved within microbial species. Engagement of innate components leads to triggering of signal pathways to promote inflammation, ensuring that invading pathogens remain in check while the specific immune response is either generated or upregulated.

ADAPTIVE IMMUNE SYSTEM

The adaptive (also called *acquired*) immune response accounts for specificity in recognition of foreign substances, or antigens, by functional receptors residing on the surface of B and T lymphocytes (Table 1-2 and Fig. 1-1). The B-cell antigen receptor (BCR) is the surface immunoglobulin, an integral glycosylated membrane protein with unique regions that bind specific antigens. There can be thousands of identical copies present on the surface of a single cell. B-cell activation occurs upon interaction of the BCR with antigen, leading to cell activation and differentiation into plasma cells, which secrete soluble immunoglobulins, or antibodies. B cells and antibodies together make up the humoral immune response. The T cell has a surface receptor structurally similar to the antibody, which also recognizes specific antigenic determinants (**epitopes**). T cells control the cellular arm of the immune response. Unlike the antibody, the T-cell receptor is present only on its surface and is not secreted. The process of T-cell activation requires a third group of cells called **antigen-presenting cells** (APCs). APCs contain surface molecules, the human leukocyte antigens, that are encoded within a gene region known as the *major histocompatibility complex*. Together these groups of molecules form a regulated pathway to present foreign antigens for subsequent recognition and triggering of specific responses to protect against disease.

BIOCHEMISTRY

Mediators of Acute Inflammation

Metabolism of phospholipids is required for production of prostaglandins and leukotrienes, both of which enhance inflammatory response by promoting increased vascular permeability and vasodilation. Prostaglandins are 20-carbon fatty acid derivatives containing a cyclopentane ring and an oxygen-containing functional group; leukotrienes are 20-carbon fatty acid derivatives containing three conjugated double bonds and hydroxyl groups.

TABLE 1-1. Innate Defensive Components

COMPONENT	EFFECTORS	FUNCTION
Anatomic and physiologic barriers	Skin and mucous membranes	Physical barriers to limit entry, spread, and replication of pathogens
	Temperature, acidic pH, lactic acid	
	Chemical mediators	
Inflammatory mediators	Complement	Direct lysis of pathogen or infected cells
	Cytokines and interferons	Activation of other immune components
	Lysozymes	Bacterial cell wall destruction
	Acute-phase proteins and lactoferrin	Mediation of response
	Leukotrienes and prostaglandins	Vasodilation and increased vascular permeability
Cellular components	Polymorphonuclear cells • Neutrophils, eosinophils • Basophils, mast cells	Phagocytosis and intracellular destruction of microorganisms
	Phagocytic-endocytic cells • Monocytes and macrophages • Dendritic cells	Presentation of foreign antigen to lymphocytes

TABLE 1-2. Key Elements of the Innate and Adaptive Immune Systems

INNATE	ADAPTIVE (ACQUIRED)
Rapid response (minutes to hours)	Slow response (days to weeks)
PMNs and phagocytes	B cells and T cells
Preformed effectors with limited variability	B-cell and T-cell receptors with a diverse array of highly selective specificities
Pattern recognition molecules recognizing structural motifs	Antibodies (humoral response)
Soluble activators	Cytokines (cellular response)
Proinflammatory mediators	
Nonspecific	Specific
No memory, no increase in response upon secondary exposure	Memory, maturation of secondary response

PMNs, polymorphonuclear neutrophils.

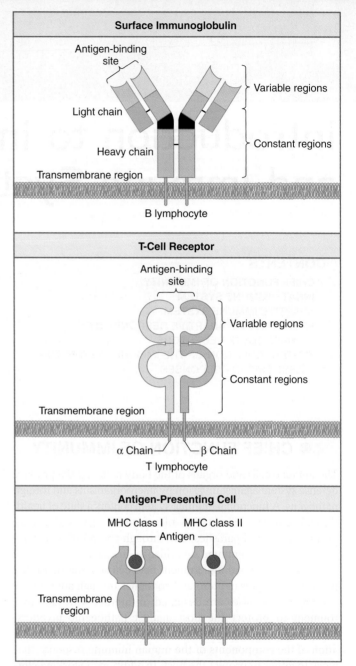

Figure 1-1. Basic structure of antigen receptors on the surface of a B cell (the immunoglobulin B-cell receptor), a T cell (the T-cell receptor), and major histocompatibility complex (*MHC*) molecules.

●●● SPECIFICITY OF ADAPTIVE RESPONSE BY LYMPHOCYTE RECEPTORS

Each B and T lymphocyte expresses a unique antigen receptor. The generation of antigen-binding specificity occurs before antigen exposure through a DNA rearrangement process that creates receptors of high diversity and binding potential. During an active immune response, a small number of B- and T-cell clones bind to the antigen with high affinity and then undergo activation, proliferation, and differentiation (Fig. 1-2). This process is called *clonal selection* and leads both to the

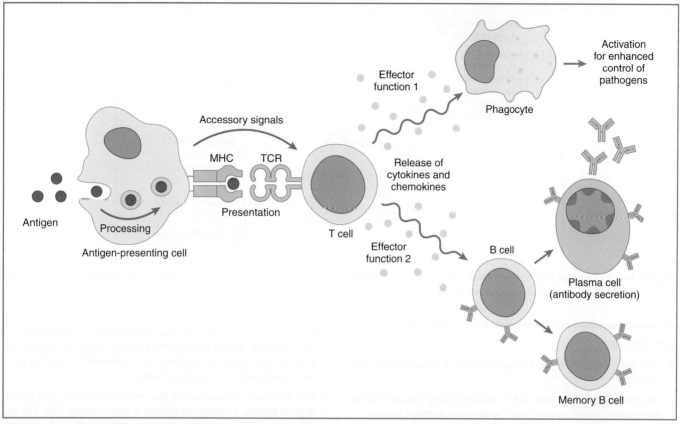

Figure 1-2. Cellular interactions drive adaptive immune functions. Antigen-presenting cells present foreign substances to activate T cells. T cells differentiate to become effectors to help phagocytes control pathogen infection or to assist B cells in production and secretion of immunoglobulins. *MHC*, major histocompatability complex; *TCR*, T-cell receptor.

production of multiple cells all with the same antigen recognition capability and to the generation of immunologic memory.

Long-term good health requires continued discrimination against foreign agents and depends on immunologic memory, which allows the adaptive immune system to respond more efficiently to previously encountered antigens (Fig. 1-3). The specific adaptive response against an antigen is much greater during secondary exposure. This principle accounts for the clinical utility of vaccines, which have done more to improve mortality rates worldwide than any other medical discovery in recorded history.

●●● TIGHT REGULATION OF THE IMMUNE SYSTEM AND ASSOCIATED RESPONSES

Immunologic diseases can be grouped into two large categories: deficiency and dysfunction (Fig. 1-4). Immunodeficiency diseases occur as the result of the absence of one or more elements of the immune system; this can either be congenital or acquired after birth. Immune dysfunction occurs when a particular immune response develops that is detrimental to the host. This deleterious response may be against a foreign antigen or a self-antigen. It may also be an inappropriate regulation of an effector response that serves to prevent a protective response. Notwithstanding the cause, the host is adversely affected. A healthy immune system occurs as a result of balance between innate and adaptive immunity,

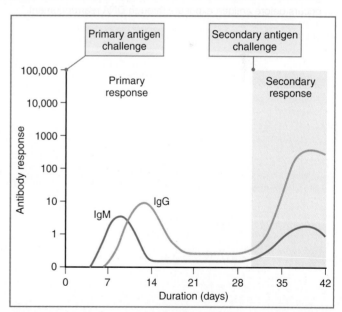

Figure 1-3. Primary and secondary antibody responses. The adaptive immune system has memory, allowing for maturation of a rapid secondary immune response with higher specificity and magnitude directed against foreign substances.

cellular and humoral immunity (see Chapters 3 and 4), inflammatory and regulatory networks (see Chapter 6), and small biochemical mediators (cytokines) (see Chapter 7). Disease occurs when the balance is altered by either deficiency or dysfunction.

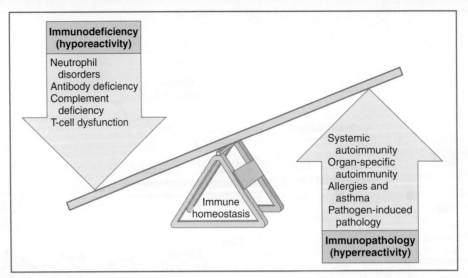

Figure 1-4. Immunodeficiency and dysfunction. Immune-based diseases can be caused by lack of specific functions (immunodeficiency) or by excessive activity (hypersensitivity).

KEY POINTS

- The chief function of the immune system is to distinguish between self and nonself.
- The immune system consists of two overlapping compartments: the innate immune system and the adaptive immune system.
- The specificity of the adaptive immune system is due to antigen-specific receptors (immunoglobulins and T-cell receptors). The generation of antigen-binding diversity inherent in these receptors occurs before antigen exposure through DNA rearrangement.
- Clonal selection occurs after immune recognition of an antigen. A small number of lymphocytes bind antigen with high affinity and undergo activation, proliferation, and differentiation into plasma cells (for B cells) or activated T cells.
- The adaptive immune system has memory, meaning that the response against a foreign substance is much greater after the first exposure. Tight regulation ensures appropriate and directed activation.

Self-assessment questions can be accessed at www.StudentConsult.com.

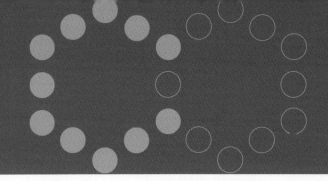

Cells and Organs of the Immune System   2

CONTENTS
PLURIPOTENT HEMATOPOIETIC STEM CELLS
MYELOID CELLS: FIRST LINE OF DEFENSE
  Neutrophils
  Eosinophils
  Basophils and Mast Cells
  Monocytes and Macrophages
  Dendritic Cells
  Platelets and Erythrocytes
LYMPHOID CELLS: SPECIFIC AND LONG-LASTING IMMUNITY
  B Lymphocytes
  T Lymphocytes
  Natural Killer T Cells
  Natural Killer Cells
PRIMARY AND SECONDARY LYMPHOID ORGANS
  Primary Lymphoid Organs
  Secondary Lymphoid Organs

The first line of defense against infection includes natural physical barriers that limit the entry of microorganisms into the body. The skin, mucosal epithelia, and cilia lining the respiratory tract represent effective mechanical barriers. Biochemical mechanisms also support innate processes to ward off potential pathogens; these include sebaceous gland secretions containing fatty acids, hydrolytic enzymes, and antibacterial defensins. Enzymes in saliva, intestinal secretions that are capable of digesting bacterial cell walls, and the acidic pH of the stomach lumen all represent innate barriers to infection. Once a pathogen has compromised these barriers and gained access to the body, cellular components must be evoked to combat invading organisms.

●●● PLURIPOTENT HEMATOPOIETIC STEM CELLS

Immune system cells are derived from pluripotent hematopoietic stem cells in the bone marrow. These cells are functionally grouped into two major categories of immune response: **innate** (natural) and **acquired** (adaptive). Innate immunity is present from birth and consists of nonspecific components. Acquired immunity by definition requires recognition specificity to foreign (*nonself*) substances (*antigens*). The major properties of the acquired immune response are specificity, memory, adaptiveness, and discrimination between self and nonself.

The acquired immune response is further classified as **humoral** or **cellular** immunity, based on participation of two major cell types. Humoral immunity involves B lymphocytes that synthesize and secrete antibodies to neutralize pathogens and toxins. Cell-mediated immunity (CMI) involves effector T lymphocytes, which can lyse infected target cells or secrete immunoregulatory factors following interaction with antigen-presenting cells (APCs) to combat intracellular viruses and organisms.

An intricate communication system allows components of innate and acquired immunity to work in concert to combat infectious disease. **Leukocytes** provide either innate or specific adaptive immunity and are derived from **myeloid** or **lymphoid** lineage (Fig. 2-1). The production of leukocytes is induced by hematopoietic growth factor glycoproteins that exert critical regulatory functions in the processes of proliferation, differentiation, and functional activation of hematopoietic progenitors and mature blood cells. Myeloid cells include highly phagocytic, motile polymorphonuclear neutrophils, monocytes, and macrophages (tissue-resident monocytes) that provide a first line of defense against most pathogens. The other myeloid cells, including polymorphonuclear eosinophils, basophils, and their tissue counterparts—mast cells—are involved in defense against parasites and in the genesis of allergic reactions. Cells from the lymphoid lineage are responsible for humoral immunity (B lymphocytes) and CMI (T lymphocytes) (Table 2-1).

HISTOLOGY

Pluripotent Hemopoietic Stem Cells

Pluripotent hemopoietic stem cells differentiate into myeloid and lymphoid lineages. The myeloid lineage gives rise to histologically distinct neutrophils with a characteristic multilobed nucleus, basophils with azurophilic granules and an S-shaped nucleus, and eosinophils with red-orange specific granules and a bilobed nucleus as well as to erythrocytes, monocytes, and platelets. Lymphoid lineages give rise to B and T lymphocytes.

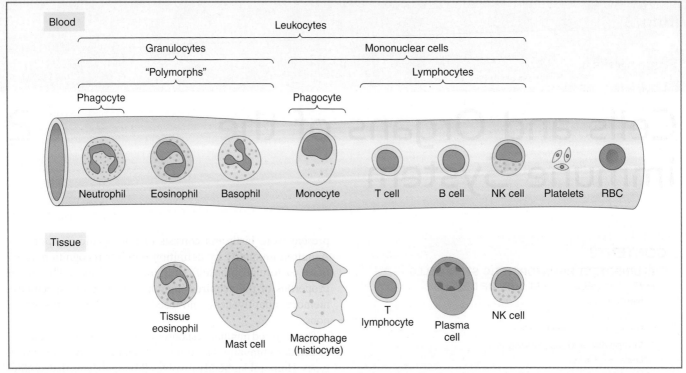

Figure 2-1. Nomenclature of immune system cells. *NK*, natural killer; *RBC*, red blood cell.

TABLE 2-1. Myeloid Leukocytes and Their Properties

PHENOTYPE	MORPHOLOGY	CIRCULATING DIFFERENTIAL COUNT*	EFFECTOR FUNCTION
Neutrophil	PMN granulocyte	2 to 7.5×10^9/L	Phagocytosis and digestion of microbes
Eosinophil	PMN granulocyte	0.04 to 0.44×10^9/L	Immediate hypersensitivity (allergic) reactions, defense against helminths
Basophil	PMN granulocyte	0 to 0.1×10^9/L	Immediate hypersensitivity (allergic) reactions
Mast cell	PMN granulocyte	Tissue specific	Immediate hypersensitivity (allergic) reactions
Monocyte	Monocytic	0.2 to 0.8×10^9/L	Circulating macrophage precursor
Macrophage	Monocytic	Tissue specific	Phagocytosis and digestion of microbes, antigen presentation to T cells
Dendritic cell	Monocytic	Tissue specific	Antigen presentation to naïve T cells, initiation of adaptive responses

PMN, polymorphonuclear.
*Normal range for 95% of population ±2 standard deviations.

●●● MYELOID CELLS: FIRST LINE OF DEFENSE

Neutrophils

Granulocyte neutrophils are the most highly abundant myeloid cell type, comprising 40% to 70% of total white blood cells. They are motile phagocytic leukocytes that are the first cells recruited to acute inflammatory sites. Neutrophils are short lived and are produced within the bone marrow through stimulation with granulocyte colony-stimulating factor. They ingest, kill, and digest microbial pathogens, with their functions dependent upon special proteins found in primary granules (containing cationic defensins and myeloperoxidase) and secondary granules (iron chelators, lactoferrin, and digestive enzymes). Neutrophilic granules contain multiple antimicrobial agents, including oxygen-independent lysozyme (peptidoglycan degradation) and lactoferrin (iron chelator). In addition, respiratory burst and granule oxidases can reduce molecular oxygen to superoxide radicals and reactive oxygen species to produce toxic metabolites (hydrogen peroxide) that limit bacterial growth.

Neutrophils dying at the site of infection contribute to the formation of the whitish exudate called **pus**.

Eosinophils

Eosinophils are polymorphonuclear granulocytes that defend against parasites and participate in hypersensitivity reactions via antibody-dependent, cell-mediated cytotoxicity mechanisms. Their cytotoxicity is mediated by large cytoplasmic granules, which contain eosinophilic basic and cationic proteins. Small granules within the cellular cytoplasm contain chemical mediators such as histamine and eosinophil peroxidase, deoxyribonuclease, ribonuclease, lipase, and major basic protein. Eosinophils are involved in manifestation of allergy and asthma via low affinity immunoglobulin E (IgE) receptors for immunoglobulins with specificity for allergen antigens. Eosinophils are specifically geared to combat multicellular parasites.

Basophils and Mast Cells

Basophils and their tissue counterpart, mast cells, are polymorphonuclear granulocytes that produce cytokines in defense against parasites. These cells are also responsible for allergic inflammation. Basophils and mast cells display high-affinity surface membrane receptors for IgE antibodies. These cells degranulate when cell-bound IgE antibodies are cross-linked by antigens secreting low-molecular-weight mediators that regulate vascular tone and capillary permeability from both primary (histamine, serotonin, and platelet-activating factor) and secondary granules (leukotrienes C_4, D_4, and B_4; prostaglandin D_2; and bradykinin). Mast cells arise from independent myeloid progenitor cells and also regulate IgE-mediated hypersensitivity responses via high-affinity IgE receptors.

Monocytes and Macrophages

Monocytes and macrophages are involved in phagocytosis and intracellular killing of microorganisms. Monocytes comprise up to 10% of circulating white blood cells. Macrophages are terminally differentiated, long-lived monocytes residing in reticular connective tissue that comprises the **reticuloendothelial system** (RES; also referred to as the *mononuclear phagocytic system* or *lymphoreticular system*). Monocytes and macrophages are motile, yet become highly adherent upon phagocytic activity. They provide natural immunity against microorganisms by a coupled process of phagocytosis and intracellular killing, recruiting other inflammatory cells through the production of cytokines and chemokines, and presenting peptide antigens to T lymphocytes for the production of antigen-specific immunity. The cells of the RES include circulating monocytes as well as tissue-resident macrophages in the spleen, lymph nodes, thymus, submucosal tissues of the respiratory and alimentary tracts, bone marrow, and connective tissues. Specialized macrophages include Kupffer cells in the liver, Langerhans cells in skin, and glial cells in the central nervous system.

> **BIOCHEMISTRY**
>
> **Vasoactive Mediators**
>
> A hallmark of inflammation includes the four cardinal signs of *tumor* (swelling), *rubor* (redness), *calor* (heat), and *dolor* (pain), resulting from biochemical actions of the vasoactive mediators prostaglandins and leukotrienes, produced from arachidonic acid precursors, and bradykinin, a peptide of the kinin group of proteins formed in response to activation of factor XII (Hageman factor). In general, these potent vasodilators cause contraction of nonvascular smooth muscle, increased vascular permeability, and pain.

Dendritic Cells

Dendritic cells (DCs) are bone marrow-derived differentiated macrophages that act as APCs to activate helper T cells and cytotoxic T cells as well as B cells. They are found in epithelia and in most organs and are important initiators for adaptive immune recognition of foreign (nonself) proteins. Subsets of DCs exist, characterized according to their location and immunological function. For example, monocytoid DCs are classical immunosurveillance cells that endocytose and enzymatically digest antigen to subsequently present to adaptive T lymphocytes. Plasmacytoid DCs are type I interferon-producing cells important in antiviral responses.

Platelets and Erythrocytes

Platelets and erythrocytes (red blood cells) arise from myeloid megakaryocyte precursors. Platelets and erythrocytes are involved in blood clotting and release inflammatory mediators involved in innate immune activation.

KEY POINTS ABOUT MYELOID CELLS

- Myeloid cells are from pluripotent hematopoietic stem cells in the bone marrow and represent the first line of defense against invading pathogens.
- Neutrophils are the most abundant of the myeloid populations. They phagocytose pathogens and fight infections using primary and secondary granules containing enzymes and molecules regulating reactive oxygen-mediated defense mechanisms.
- Eosinophils are polymorphic granulocytes that are critical for defense against large multicellular pathogens.
- Basophils and mast cells are the least common of the granulocytes. They are important in asthma and allergic responses.
- DCs, monocytes, and macrophages are critical for mediation of proinflammatory function and are important APCs that can show foreign proteins to adaptive lymphocyte populations.

LYMPHOID CELLS: SPECIFIC AND LONG-LASTING IMMUNITY

Lymphoid cells are present in healthy adults at concentrations of approximately 1.3 to 3.5×10^9/L. Lymphocytes differentiate into three separate lines: (1) thymic-dependent cells or T lymphocytes that operate in cellular and humoral immunity, (2) B lymphocytes that differentiate into plasma cells to secrete antibodies, and (3) natural killer (NK) cells that can lyse infected target cells. T and B lymphocytes produce and express specific receptors for antigens whereas NK cells do not (Table 2-2).

B Lymphocytes

B lymphocytes differentiate into **plasma cells** to secrete antibodies, which are Ig glycoproteins that bind antigens with a high degree of specificity. The genesis of mature B cells from pluripotent progenitor stem cells occurs in the bone marrow and is antigen independent. The activation of B cells into antibody-producing and antibody-secreting cells (plasma cells) is antigen dependent. Mature B cells can have 1 to 1.5×10^5 immunoglobulin receptors for antigen embedded within their plasma membrane. Once specific antigen binds to surface Ig molecule, the B cells differentiate into plasma cells that produce and secrete these antibodies. Because all the antibodies produced by the individual plasma cell have the same antigenic specificity, the antibodies are referred to as *monoclonal*. If B cells also interact with T helper cells, they proliferate and switch the isotype (class) of immunoglobulin that is produced while retaining the same antigen-binding specificity. T helper cells are thought to be required for switching from IgM to IgG, IgA, or IgE isotypes. B cells also process and present protein antigens. In this case, immunoglobulin receptors recognize and internalize antigen, which then is degraded and presented to T lymphocytes.

T Lymphocytes

Immature thymocytes differentiate in the thymus, where rearrangement of antigen-specific T-cell receptor (TCR) genes gives rise to a diverse set of clonotypic T lymphocytes. In the thymus, cells are selected for maturation only if their TCRs do not interact with self-peptides presented in the context of self–major histocompatibility complex (MHC) molecules on APCs. T lymphocytes are involved in the regulation of the immune response and in CMI and help B cells produce antibody. Mature lymphocytes also display one of two accessory molecules—CD4 or CD8—that define whether a T cell will be a CD4-expressing **helper T lymphocyte** or a CD8-expressing **cytotoxic T lymphocyte** (CTL). Every T lymphocyte also expresses CD3, a multisubunit cell-signaling complex noncovalently associated with the antigen-specific TCR. The TCR/CD3 complex recognizes antigens associated with the MHC molecules on APCs or target cells (e.g., virus-infected cells). Upon recognition of presented antigen, the T lymphocyte becomes activated and secretes cytokines (interleukins).

> **HISTOLOGY**
>
> **Cluster of Differentiation**
>
> CD (cluster of differentiation) designates cell surface proteins. Each unique molecule is assigned a different number designation. Surface expression of a particular CD molecule may not be specific for just one cell or even a cell lineage; however, many are useful for characterization of cell phenotypes.

T Helper Lymphocytes

T helper lymphocytes are the primary regulators of delayed-type hypersensitivity responses. They express the CD4 molecule and regulate antigen-directed effector functions involved in CMI to pathogens. They also assist in the stimulation of B lymphocytes to proliferate and differentiate to become antibody-producing cells. T helper lymphocytes recognize foreign antigen complexed with MHC class II molecules on DCs, B cells, and macrophages or other APCs that can express MHC class II.

Cytotoxic T Lymphocytes

CTLs are cytotoxic against tumor cells and host cells infected with intracellular pathogens. These cells usually express CD8 and destroy infected cells in an antigen-specific manner that is

TABLE 2-2. Lymphoid Leukocytes and Their Properties

TOTAL LYMPHOCYTES		1.3 TO 3.5×10^9/L	EFFECTOR FUNCTION
B cell	Monocytic	Adaptive	Humoral immunity
Plasma cell	Monocytic	Adaptive	Terminally differentiated, antibody-secreting B cell
T cell	Monocytic	Adaptive	Cell-mediated immunity, immune response regulation
NKT cell	Monocytic	Adaptive	Cell-mediated immunity (glycolipids)
NK cell	Monocytic	Innate	Innate response to microbial or viral infection

NK, natural killer; *NKT cell*, natural killer T cell.

dependent upon the expression of MHC class I molecules that are expressed on almost all nucleated cells in the body. **T suppressor cells** express CD8 molecules and are thought to be related to CTLs. The suppressor cells function to suppress and limit T- and B-lymphocyte–specific responses.

Natural Killer T Cells

NK T cells (NKTs) are a heterogeneous group of T cells that share properties of both T cells and NK cells. They recognize foreign lipids and glycolipids and constitute only 0.2% of all peripheral blood T cells. It is now generally accepted that the term *NKT cells* refers to a restricted population of T cells coexpressing a heavily biased, semi-invariant TCR and NK cell markers. NKT cells should not be confused with NK cells.

Natural Killer Cells

NK cells are large granular lymphocytes that kill tumor cells and virus-infected targets. NK cells do not express antigen-specific receptors such as the TCR/CD3 complex. Instead, NK cells express a variety of killer immunoglobulin-like receptors, which can bind MHC class I molecules and stress molecules on target cells and send either a positive or negative signal for NK cell activation. NK cells adhere to infected target cells and induce their cell death via delivery of apoptic signals mediated by perforins, granzymes, and tumor necrosis factor-α, or by effector interactions via their surface Fas ligand molecule with Fas on the target cell. NK cells can also kill through antibody-dependent, cell-mediated cytotoxicity mechanisms via cell surface receptors for constant domains present on immunoglobulins.

KEY POINTS ABOUT LYMPHOID CELLS

- Lymphoid cells are from pluripotent hematopoietic stem cells in the bone marrow and represent the secondary line of defense against invading pathogens.
- B lymphocytes make antibodies that specifically recognize antigenic determinants. Activated B lymphocytes are called *plasma cells*.
- T lymphocytes are involved in cell-mediated immune function. They are subdivided into helper and cytotoxic populations.
- NKT cells recognize foreign glycolipids. They share properties of both T cells and NK cells.
- NK cells do not have a specific antigen receptor but do have the ability to kill tumor and virally infected cells.

●●● PRIMARY AND SECONDARY LYMPHOID ORGANS

The lymphatic organs are tissues in which leukocytes of myeloid and lymphoid origin mature, differentiate, and proliferate (Fig. 2-2). Lymphoid organs are composed of epithelial and stromal cells arranged either into discretely capsulated organs or accumulations of diffuse lymphoid tissue. The **primary**

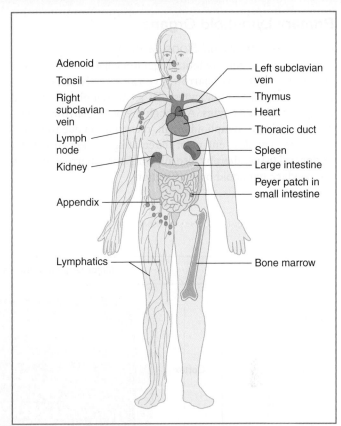

Figure 2-2. Distribution of lymphoid tissues.

(central) lymphoid organs are the major sites of lymphopoiesis, for example, where B and T lymphocytes differentiate from stem cells into mature antigen-recognizing cells. The **secondary lymphoid organs** are those tissues in which antigen-driven proliferation and differentiation occur.

Historically, the primary lymphoid organ was discovered in birds, in which B lymphocytes undergo maturation in the bursa of Fabricius, an organ situated near the cloaca. Humans do not have a cloaca, nor do they possess a bursa of Fabricius. In embryonic life, B lymphocytes mature and differentiate from hematopoietic stem cells in the fetal liver. After birth, B cells differentiate in the bone marrow. Maturation of T lymphocytes occurs in a different manner. Progenitor cells from the bone marrow migrate to the thymus, where they differentiate into T lymphocytes. The T lymphocytes continue to differentiate after leaving the thymus, and are driven to do so by encounter with APCs presenting trapped circulating antigen in the secondary lymphoid organs.

ANATOMY AND EMBRYOLOGY

Ontogenic Development of Lymphoid Cells

Lymphocyte stem cells are produced first by the omentum and later by the yolk sac or fetal liver. In older fetuses and adults, the bone marrow is the major source of lymphocytes derived from pluripotent hematopoietic stem cells.

Primary Lymphoid Organs

Fetal Liver and Adult Bone Marrow

Islands or foci of hemopoietic progenitor cells in the fetal liver and in the adult bone marrow give rise directly to polymorphonuclear cells, monocytes, DCs, B lymphocytes, and precursor T lymphocytes. Although the bone marrow is technically a primary lymphoid organ, recirculation due to vascularization enables entry of circulating leukocytes from peripheral tissue, thereby also allowing bone marrow to serve a secondary lymphoid organ function.

Thymus Gland

The principal function of the thymus gland is to educate T lymphocytes to differentiate between self and nonself antigens. The thymus reaches maturity before puberty and slowly loses function thereafter. This lymphoepithelial organ is primarily composed of stroma (thymic epithelium) and cells of the T-lymphocyte lineage (Fig. 2-3). The thymus is divided into lobules containing cortex and medulla regions. Precursor T lymphocytes (**thymocytes**) differentiate to express specific receptors for antigen. The cortex contains immature thymocytes with few macrophages. The maturing lymphocytes pass

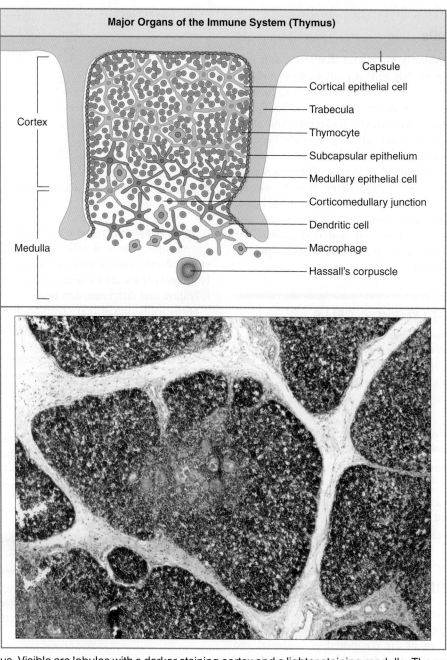

Figure 2-3. The thymus. Visible are lobules with a darker staining cortex and a lighter staining medulla. The medulla is characterized by the presence of Hassall's corpuscles.

through the medulla, interacting with epithelial cells, DCs, and macrophages. Only 5% to 10% of thymocytes leave the thymus for final maturation in secondary organs. Hassall corpuscles are a characteristic morphologic feature located within the medullary region of the thymus.

Secondary Lymphoid Organs

The secondary lymphoid organs provide localized environments where lymphocytes may respond to pathogens and foreign antigens. The spleen and lymph nodes are the major secondary lymphoid organs. Additional secondary lymphoid organs include the mucosa-associated lymphoid tissue (MALT), which is composed of cellular aggregates in the lamina propria of the digestive tract lining and respiratory tract.

Spleen

The spleen is a filter for blood that is histologically composed of two tissue types, red pulp and white pulp. The red pulp is made up of vascular sinusoids containing large numbers of macrophages, and is actively involved in the removal of dying and dead erythrocytes and of infectious agents (Fig. 2-4). The white pulp contains the lymphoid tissue, which is arranged around a central arteriole as a periarteriolar lymphoid sheath and is composed of T- and B-cell areas and follicles containing germinal centers. Dendritic reticular cells and phagocytic

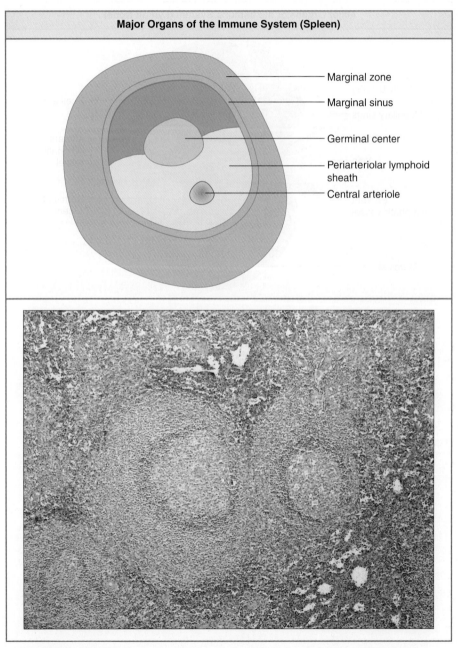

Figure 2-4. The spleen. The white pulp of the spleen contains a central artery and associated follicle (germinal center, marginal zone, and periarteriolar lymphoid sheath).

macrophages found in germinal centers present antigen to lymphocytes. The germinal centers are where B cells are stimulated to become plasma cells that produce and secrete antibodies.

Lymph Node

Lymph nodes (Fig. 2-5) form part of the lymphatic network that filters antigen and debris from lymph during its passage from the periphery to the thoracic duct. The **lymphatic vessels** (or lymphatics) are a network of thin tubes that branch into tissues throughout the body. Lymphatic vessels carry a colorless, watery fluid called **lymph**, which originates from interstitial fluid. Plasma and leukocytes from vascular tissue and capillary beds continuously circulate via afferent lymphatics, eventually arriving at draining lymph nodes. Histologically, the lymph node is composed of an outer sinuous connective capsule. Subcapsular afferent vessels deliver antigen-containing fluid and draining pathogens into discrete lobules, where they are phagocytosed by antigen-presenting macrophages and DCs. Naive T lymphocytes enter the lymph node through specialized high endothelial venules and travel to the paracortex, where they encounter the APCs. A B-lymphocyte–rich cortex contains both primary and secondary follicles and active germinal centers representing regions of lymphocyte division, differentiation, and expansion. Activated and memory lymphocytes eventually migrate through the central medullary sinus, leaving through efferent lymphatic vessels.

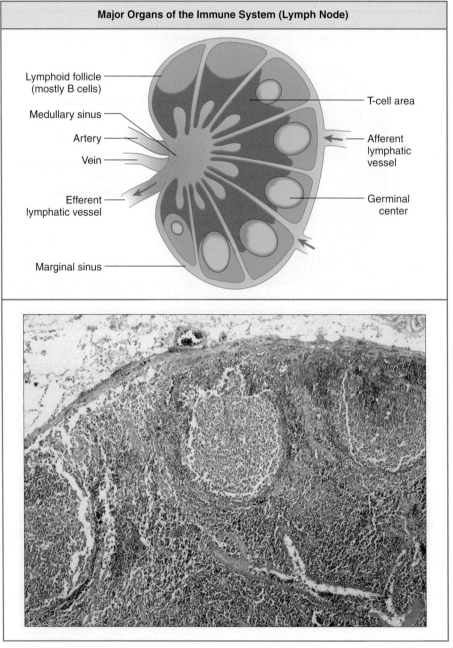

Figure 2-5. Lymph node. The normal architecture of a lymph node containing afferent and efferent lymphatic vessels, T-cell–rich paracortex, and germinal centers of activation.

Mucosa-Associated Lymphoid Tissue

Multiple nodules of partially encapsulated lymphoid tissue and loosely associated lymphoid aggregates are found in epithelia and lamina propria of mucosal surfaces (Fig. 2-6). Collectively, these aggregates located along the gastrointestinal and respiratory tracts are called MALT, which can be further classified as gut-associated lymphoid tissue (GALT) or bronchus-associated lymphoid tissue (BALT). The tonsils, appendix, and Peyer patches are representative of lymphoid tissue found in and around mucosal epithelia. In particular, specialized microfold (M) cells of the follicle-associated epithelium of the MALT in gut and the respiratory system play a critical role in the genesis of immune responses by delivering foreign (nonhost) material via transcytosis to underlying lymphoid tissue. Specialized resident intraepithelial lymphocytes with potent cytolytic and immunoregulatory capacities monitor mucosal tissue to help defend against pathogenic infection.

KEY POINTS ABOUT LYMPHOID ORGANS

- The thymus and bone marrow represent primary lymphoid organs; they are the sites of lymphopoiesis where cells develop and learn to differentiate self from nonself.
- The spleen is a filter for blood and is composed of parenchyma that allows both innate and adaptive cells to interact and respond to presented antigens.
- The lymph nodes are connected via lymphatics and represent a local environment for antigen drainage and interaction with presenting populations to engage adaptive lymphocytes.
- MALT represents cellular aggregates loosely associated with the epithelia and lamina propria of mucosal parenchyma, delivering foreign materials to underlying lymphocytes. The tonsils, appendix, and Peyer patches represent a more formal association of accumulated tissue that shares this same functional parameter.

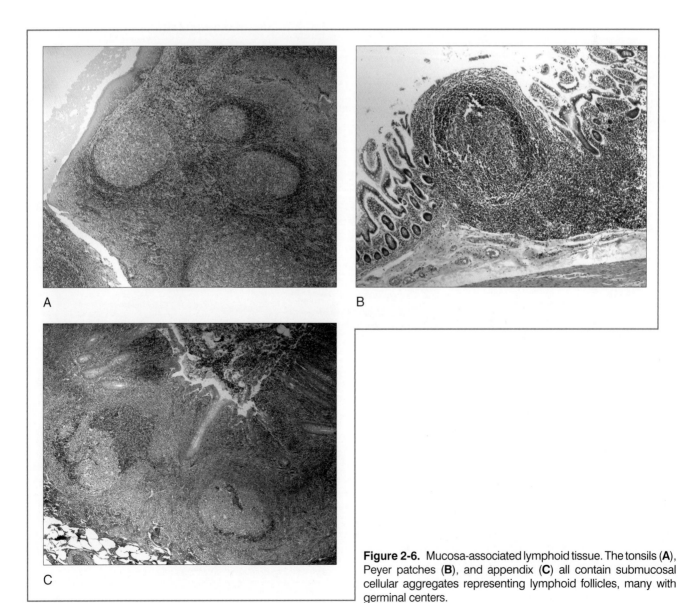

Figure 2-6. Mucosa-associated lymphoid tissue. The tonsils (**A**), Peyer patches (**B**), and appendix (**C**) all contain submucosal cellular aggregates representing lymphoid follicles, many with germinal centers.

KEY POINTS

- Physical barriers and biochemical mechanisms comprise the first line of defense against foreign pathogens. Once inside the body, innate and cellular components are required to fight against infectious agents.
- Innate cells and components are present from birth and represent a nonspecific first line of defense to foreign substances.
- Stem cell progenitors within the bone marrow are the precursors to myeloid and lymphoid progenitors, giving rise to cells involved in either innate or adaptive immune responses. Myeloid cells include neutrophils, eosinophils, basophils, mast cells, monocytes, macrophages, and DCs. Lymphoid progenitors give rise to B and T lymphocytes, NKT cells, and NK cells.
- Acquired (adaptive) immune responses discriminate between self and nonself and demonstrate specificity, memory, and adaptiveness. Humoral immunity refers to B lymphocytes, which produce antibodies that neutralize pathogens and toxins. Cellular immunity encompasses T lymphocytes to eradicate intracellular organisms.
- Primary lymphoid organs, such as the bone marrow and thymus, are the major sites of lymphopoiesis where lymphocytes differentiate. Secondary lymphoid organs, such as the spleen and lymph nodes, are locations within the body where antigen-driven proliferation and maturation of lymphocytes occur. The MALT represents a loosely associated lymphoid aggregate where APCs located in the mucosal epithelia present antigens to lymphocytes.

Self-assessment questions can be accessed at www.StudentConsult.com.

Humoral Immunity: Antibody Recognition of Antigen

3

CONTENTS
ANTIBODY STRUCTURE AND FUNCTION
ANTIBODY-ANTIGEN INTERACTIONS
   Antigen Binding
   Physiochemical Forces in Antigen-Antibody Interactions
ANTIGENS AND IMMUNOGENS
   Recognition of Sequential and Conformational Epitopes
   Cross-Reactivity
GENETIC BASIS OF ANTIBODY STRUCTURE
GENERATION OF ANTIBODY DIVERSITY
   Multiple Variable Gene Segments
   Combinatorial Diversity
   Heavy and Light Chain Combinations
   Junctional and Insertional Diversity
ISOTYPE SWITCHING AND AFFINITY MATURATIONS

Humoral immunity refers to the arm of the acquired immune system that is mediated by antibody (immunoglobulin) recognition of antigens associated with foreign substances or pathogens. Antigen recognition is coupled with the ability to initiate biologic responses that protect against microorganisms and neutralize viruses. The specific recognition of foreign antigen by B lymphocytes occurs through membrane-bound receptors and triggers proliferation and differentiation into antibody-producing plasma cells.

●●● ANTIBODY STRUCTURE AND FUNCTION

The antibody is a tetrameric polypeptide structure with distinct biologic activity attributed to each end of the molecule (Fig. 3-1). Antibodies are composed of two identical heavy chain and two identical light chain polypeptides. Both heavy and light chain molecules have variable and constant domains and interact via intradisulfide and interdisulfide linkages. The variable region, termed $F(ab')_2$ (fragment, antibody binding), confers antigen recognition. The constant region, termed Fc (fragment, crystalline), interacts with cell surface receptors. The heavy chain contains a hinge domain that confers flexibility to allow optimal binding to antigen.

The constant regions of the heavy chain confer antibody function representing five different classes, or **isotypes** (Fig. 3-2). Each class of heavy chain has a characteristic amino acid sequence that distinguishes it from the other four classes, but all five classes have a significant percentage of amino acid sequence similarities. The isotypes, immunoglobulin (Ig) M, IgD, IgG, IgE, and IgA, have characteristic properties (Table 3-1). The different heavy chains corresponding to their class are given Greek letter designations: γ, α, μ, ϵ, and δ. In many species, there are two or more subclasses of some heavy chains that differ from one another by only a few amino acids; humans have nine possible heavy chains each with unique biologic functions. There are four subclasses of the IgG isotype, called IgG_1, IgG_2, IgG_3, and IgG_4, as listed in Table 3-2. IgA has two subclasses called IgA_1 and IgA_2.

Light chains come in two varieties, called kappa (κ) and lambda (λ). The difference between the two types of light chains is in the amino acid sequence of the constant region domain. The overall ratio of the two light chain types in human immunoglobulin is approximately 60% κ and 40% λ.

Both heavy and light chains have functional domains—amino acid sequences giving regularity to structure by way of disulfide-bridged loops. There are two domains on both κ and λ light chains and either four or five domains on heavy chains. The amino acid sequences in the first domain on both light and heavy chains vary greatly from molecule to molecule and are referred to as the *variable light domain* (V_L) or *variable heavy domain* (V_H). The light chain domain, which is constant in its amino acid sequence for the κ or λ type of chain, is referred to as the C_L *domain*. The constant domains of heavy chains are numbered from the amino terminal to the carboxyl terminal as C_H1, C_H2, C_H3, and C_H4 (for IgM and IgE).

IgG, IgA, and IgD genes each have an exon coding for a short span of amino acids that occupy the space between the C_H1 and C_H2 domains. This segment is rich in cysteine and proline and permits significant flexibility between the two arms of the antibody; the area is called the **hinge region.** This stretch is highly susceptible to protease cleavage.

Vertebrates and invertebrates express a large number of closely related cell-surface proteins, many of which appear to have evolved from common gene sequences. Collectively,

Humoral Immunity: Antibody Recognition of Antigen

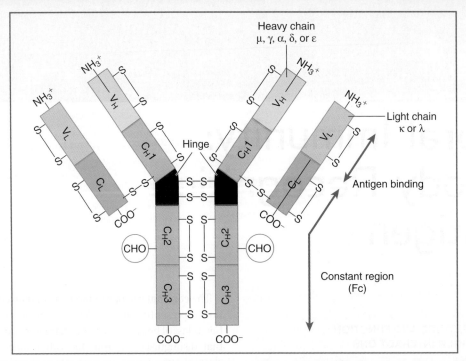

Figure 3-1. Antibody structure and functional domains. The basic structure of the antibody contains heavy chains and light chains, showing intradisulfide and interdisulfide bonds and the characteristic hinge region. The interactions between variable domains constitute the antigen-binding domain, while the constant regions confer specific biologic properties to the molecule. *CHO*, carbohydrate.

they are called the *immunoglobulin gene superfamily*. All members of the immunoglobulin gene superfamily contain Ig domains, which share the primary amino acid sequence and physical structure of β-pleated sheets with intrachain disulfide bonds. Members of the Ig gene superfamily include surface receptors and adhesion molecules.

KEY POINTS ABOUT ANTIBODY STRUCTURE AND FUNCTION

- The basic immunoglobulin (antibody) unit consists of two light and two heavy chains.
- Each set of chains are covalently linked by disulfide bridges, allowing for unique generation of antigen binding regions.
- The specific domains within the constant chains allow for unique biologic functions.
- Five subclasses of heavy chains (γ, α, μ, ε, and δ) correspond to distinct heavy chains (IgG, IgA, IgM, IgD, and IgE).
- There are two types of light chains in humans (κ and λ), which differ in amino acid content with their respective constant domains.

BIOCHEMISTRY

Amino Acid Protein Structure

The combination of disulfide links between cysteine residues and the proline-rich hinge region gives the antibody molecule the unique structure and flexibility necessary to interact with antigens.

ANTIBODY-ANTIGEN INTERACTIONS

Antigen Binding

The molecular conformation in the variable region of an antibody that confers antigenic specificity is referred to as the **idiotype**; the site of direct contact in the variable region is called the **paratope.** Each antibody molecule can bind between two and 10 antigenic determinants (as long as they are physically close together), depending on the number of available F(ab) regions present. IgG, IgD, and IgE have two F(ab) binding sites. IgA as a dimer has four binding sites; the pentameric IgM molecule has 10 binding sites. An aggregated antigen with multiple repeating epitopes (such as those found in bacterial capsules or carbohydrate ABO blood groups) allows increase in total binding strength, thereby raising the overall **avidity** (or **functional affinity**) of the interaction (Fig. 3-3).

Physiochemical Forces in Antigen-Antibody Interactions

Antibodies can bind a wide variety of molecules with high specificity, ranging from large macromolecules to small chemical moieties. The molecular region on the antigen recognized by immune components is called an **epitope,** or **antigenic determinant.** Antibody binding to antigen does not involve covalent chemical bonds. The strength of binding to epitopes on the antigen, or the interaction affinity, is based on multiple forces present within the binding site (Fig. 3-4).

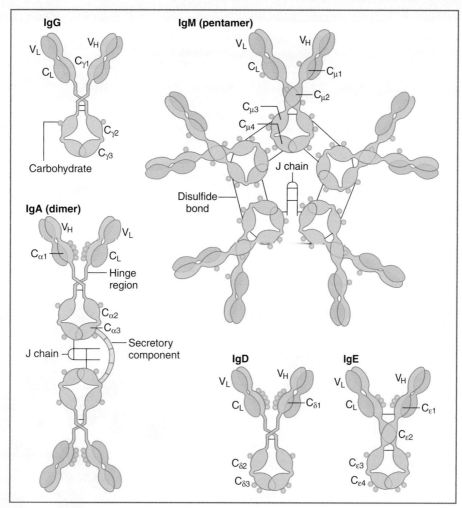

Figure 3-2. Pictured are the five classes of antibody isotypes representing monomeric immunoglobulin (*Ig*) G, IgD, and IgE, pentameric IgM, and dimeric IgA with secretory component.

TABLE 3-1. Classes of Antibody Isotypes and Functional Properties*

ISOTYPE	IMMUNOGLOBULIN CLASS				
	IgM	IgD	IgG	IgE	IgA
Structure	Pentamer	Monomer	Monomer	Monomer	Monomer, dimer
Heavy chain designation	μ	δ	γ	ε	α
Molecular weight (kDa)	970	184	146–165	188	160 × 2
Serum concentration (mg/mL)	1.5	0.03	0.5–10.0	<0.0001	0.5–3.0
Serum half-life (days)	5–10	3	7–23	2.5	6
J chain	Yes	No	No	No	Yes
Complement activation	Strong	No	Yes, except IgG$_4$	No	No
Bacterial toxin neutralization	Yes	No	Yes	No	Yes
Antiviral activity	No	No	Yes	No	Yes
Binding to mast cells and basophils	No	No	No	Yes	No
Additional properties	Effective agglutinator of particulate antigens, bacterial opsonization	Found on surface of mature B ells, signaling via cytoplasmic tail	Antibody-dependent cell cytotoxicity	Mediation of allergic response, effective against parasitic worms	Monomer in secretory fluid, active as dimer on epithelial surfaces

*Refer also to Figure 3-2.
Ig, immunoglobulin.

TABLE 3-2. Unique Biologic Properties of Human IgG Subclasses

	IgG$_1$	IgG$_2$	IgG$_3$	IgG$_4$
Occurrence (% of total IgG)	70	20	7	3
Half-life (days)	23	23	7	23
Complement binding	+	+	Strong	No
Placental passage	++	±	++	++
Receptor binding to monocytes	Strong	+	Strong	±

Ig, immunoglobulin.

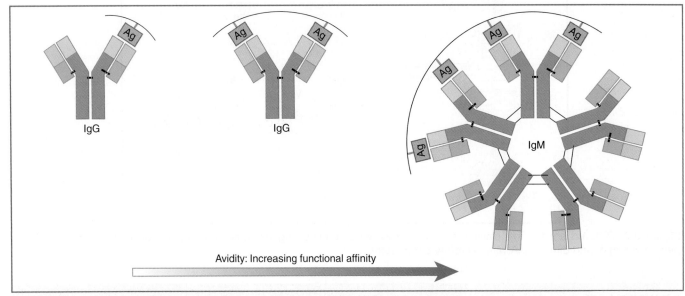

Figure 3-3. Multivalent interactions increase functional affinity of antibody-antigen binding. Interactions of repeating antigens (*Ag*) with more than one functional antigen-binding region allow greater stabilization of binding and increase functional affinity. Monomeric immunoglobulin G (*IgG*) can bind either one or two epitopes, whereas pentameric immunoglobulin M (*IgM*) can bind up to 10.

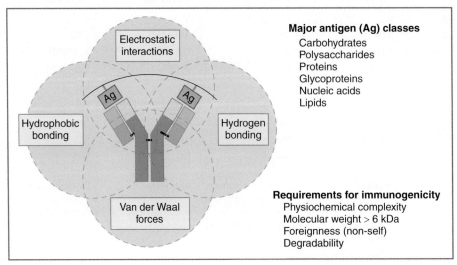

Figure 3-4. Noncovalent forces contribute to avidity of antibody-antigen interactions. Binding of antigen to antibody is a noncovalent interaction. The major forces involved are depicted.

● ● ● ANTIGENS AND IMMUNOGENS

An antigen is any substance that can be recognized by the immune system. Major classes of antigens include proteins, carbohydrates, lipids, and nucleic acids. In contrast, an **immunogen** is any substance that can evoke an immune response. Not all antigens are immunogenic. For example, **haptens** are low-molecular-weight compounds that are nonimmunogenic by themselves but are antigenic; haptens become immunogenic after conjugation to high-molecular-weight immunogenic carriers. The rule to remember is that *all immunogens are antigens, but not all antigens are immunogens.* This concept is important when considering rational design of vaccines.

BIOCHEMISTRY
Noncovalent Bond Forces

The weak interactions represented by noncovalent bond forces are important in multiple biologic systems and allow for fluidity of information, as is seen with ligands interacting with cellular receptors. One of these interactions is the van der Waals forces, which are relatively weak electric forces that attract neutral atoms and molecules to one another in gases, in liquefied and solidified gases, and in almost all organic liquids and solids. These forces arise from polarization induced in each particle by the presence of other particles.

Recognition of Sequential and Conformational Epitopes

Two general classes of epitopes can be distinguished on antigens. They are best described as they exist on protein antigens, but other classes of antigens (e.g., carbohydrates and nucleic acids) can exhibit both kinds of epitopes under some circumstances. **Sequential epitopes** are short stretches of amino acids (4 to 7 inches length) that can be recognized by antibodies when the short peptide exists free in solution or when it is chemically coupled to another protein molecule. **Conformational epitopes** require the native three-dimensional configuration of the molecule to be intact, and antigenic determinants need not be contiguous; denaturation of the molecule destroys these kinds of epitopes.

Cross-Reactivity

The forces mediating antigen-antibody recognition allow a high degree of specificity. That is, antibodies specific for one epitope or hapten can easily distinguish that epitope or hapten from other similar structures. This specificity is not absolute; antibodies specific for one epitope can bind with structurally similar nonidentical epitopes with a lower affinity. **Cross-reactivity** refers to the situation in which an antibody can react with two similar molecules because they share one or more identical epitope or the epitope in question is similar enough in sequence or shape to bind with weaker affinity. For example, antibodies elicited with toxoids react with native toxins, allowing clinical application for vaccination with nonpathogenic antigens such as tetanus toxoid and diphtheria toxoid.

KEY POINTS ABOUT ANTIBODY-ANTIGEN INTERACTIONS

- Antibodies recognize antigenic determinants (epitopes) via binding pockets found within their combined heavy and light chain sequences.
- Major classes of antigens include proteins or glycoproteins, nucleic acids, carbohydrates, and lipids.
- The binding site for antigen occurs in hypervariable regions. The binding is not a covalent interaction. Rather, multiple weak forces stabilize the binding. These forces include electrostatic interactions, hydrogen bonding, hydrophobic interactions, and van der Waals forces.
- Cross-reactivity refers to reactivity of an antibody with two similar molecules that share one or more physically similar epitopes.

MICROBIOLOGY
Antibacterial Vaccines

Group B streptococcus causes invasive infections of newborns and adults, with commonly reported bacteremia, meningitis, and pneumonia. The associated polysaccharide capsule antigens are being targeted for use in vaccines with methods that conjugate carbohydrates to a protein carrier to create T-dependent antigens.

● ● ● GENETIC BASIS OF ANTIBODY STRUCTURE

The generation of antigen-binding capability of the antibody B-cell receptor occurs before antigen exposure through DNA rearrangement involving combinations of multiple genes to achieve high diversity needed for immune responses. Recombination occurs for both heavy and light chain genes by way of an enzyme complex known as **V(D)J recombinase.** The recombinase is the product of two genes (*RAG-1* and *RAG-2*; recombinase-activating genes); defects in either *RAG-1* or *RAG-2* can cause a spectrum of severe immunodeficiencies with devastating clinical complications.

The organization of heavy chain genes shows three different regions contributing to production of the variable region. These gene segments are referred to as V_H (variable), D_H (diversity), and J_H (joining) (Fig. 3-5). There are approximately 50 V_H genes, 20 D_H genes, and up to 6 J_H genes present in germline DNA. The V(D)J recombinase mediates joining of different gene segments through mechanisms by which intervening segments are spliced out of the genome within that particular B cell. This process requires pairing of 7-base-pair and 9-base-pair gene sequences, after which the intervening DNA is "looped out" and deleted permanently from the chromosome. To complete the process, alternative splicing brings the rearranged VDJ sequence together with distinct gene segments coding for the C_H (constant) region.

Similar events occur for rearrangement of the light chain locus. In humans (and most other mammals) there are two light

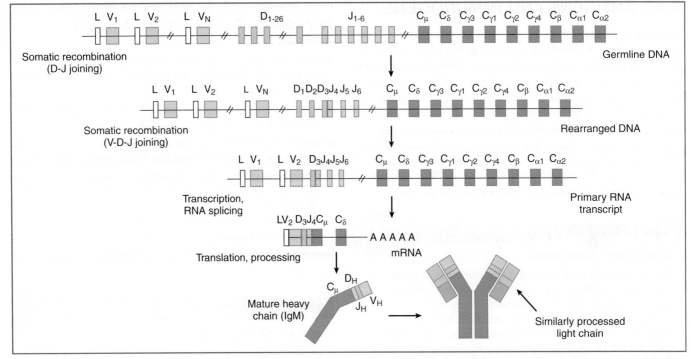

Figure 3-5. Genetic organization and recombination events. Antibody diversity is generated by DNA recombination events that randomly fuse variable, diversity, and joining regions. The recombination is accomplished in a defined order by enzymes RAG-1 and RAG-2. The first events culminate in transcribing mRNA coding for immunoglobulin M and D; differential translation determines whether mature polypeptide will be one or the other. *RAG*, recombinant activating genes; *L*, leader sequence; *V*, variable; *D*, diversity; *J*, joining; *C*, constant region.

chain loci—κ and λ—located on different chromosomes. The germline arrangement for light chains is similar to that of the heavy chain locus, except there are no D gene segments. In addition, for the κ locus there is only one constant region, whereas the λ locus has multiple constant regions, each with its own J gene segment.

> **BIOCHEMISTRY**
>
> **Gene Transcription and RNA Translation**
>
> The production of protein involves gene transcription of DNA into primary RNA transcripts by polymerases, under the control of transcription factors. The primary transcript is further processed by splicing out noncoding introns, to produce full messenger RNA (mRNA), which is eventually translated into protein. In B lymphocytes, the zinc finger transcription factor early growth response 1 (*Egr-1*) is one of the many immediate-early genes induced upon B-cell antigen receptor engagement.

●●● GENERATION OF ANTIBODY DIVERSITY

The portions of the variable region that participate in antigen-binding are called **complementarity-determining regions** (CDRs). Amino acids in these hypervariable regions contact residues for antigen; the CDRs form the region of structural complementarity for antigenic epitopes, and differences in antigen-binding are due to differences in these sequences. Two of the CDRs (CDR1 and CDR2) are "hard wired" into the V gene segment and thus depend upon the V segment selected during rearrangement. CDR3 consists of the junction of the V, D, and J gene segments and hence has a high degree of variability. The CDRs of both the heavy and light chain participate in the formation of the antigen-binding pocket (paratope).

Four major mechanisms for generating antibody diversity in humans occur during B-cell development, before antigen exposure. These mechanisms give rise to a repertoire of antibodies that in theory have the capability of recognizing approximately 10^{14} different epitopes.

Multiple Variable Gene Segments

There are more than 50 V gene segments in the heavy chain locus; there are approximately 40 V gene segments each in the κ light chain loci and nearly as many V gene segments in the λ loci.

Combinatorial Diversity

The V, D, and J regions in heavy chains (and the V and J regions in light chains) are selected randomly during V(D)J rearrangement ("joining"). Thus, 50 V_H genes × 26 D_H genes × 6 J_H genes yield more than 6000 possible heavy chain VDJ loci combinations. A lesser degree of variation occurs in the light chain recombination because there are no D regions; there are about 200 and about 160 different VJ combinations in κ and λ, respectively.

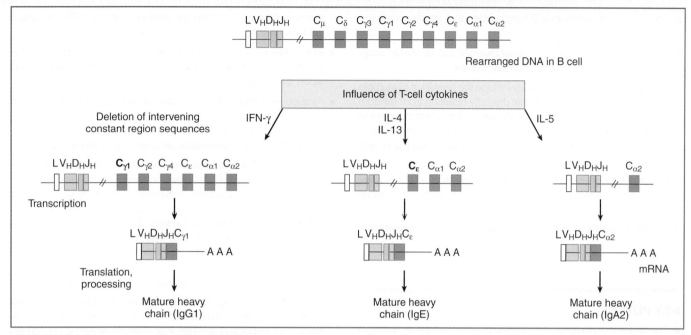

Figure 3-6. Isotype class switching. During plasma cell differentiation, the antibody isotype may be changed where the same variable region is recombined with a different constant region gene sequence. Intervening DNA is excised, allowing generation of mRNA for different isotypes. Subsequent switching may occur to any downstream remaining coding sequence; once done, however, the change is irreversible. *INF,* interferon; *Ig,* immunoglobulin; *L,* leader sequence; *V,* variable; *D,* diversity; *J,* joining; *C,* constant region.

Heavy and Light Chain Combinations

Because the heavy and light chain loci recombine independently, each B cell will contain a different combination of randomly assorted heavy and light chains.

Junctional and Insertional Diversity

The recombination between V and D-J segments and between D and J segments is not precise, often staggering the junctional location between recombined segments by a few base pairs. This "sloppiness" causes differences in the amino acid sequence and leads to junctional diversity. Insertional diversity results from the activity of terminal deoxynucleotide transferase, an enzyme that is expressed during heavy chain rearrangement. This enzyme adds nucleotides randomly at the V-D and D-J junctions.

●●● ISOTYPE SWITCHING AND AFFINITY MATURATION

Secretion of antibodies occurs only after antigenic stimulation of membrane-anchored immunoglobulin on the B-cell surface. Differentiation into plasma cells usually occurs in germinal centers of lymph nodes, the spleen, or mucosa-associated lymphoid tissue. The process requires both antigenic recognition and help given in the form of T-cell cytokines. During differentiation, two processes can occur to change the biologic properties of the secreted antibody. The first is class switching, or **isotype switching,** in which deletion of intervening DNA sequences occurs, allowing antibodies to switch from IgM and IgD to another isotype (Fig. 3-6).

Isotype switching results when antigen-stimulated B cells receive a cytokine signal from T helper cells. The V region does not change during isotype switching; therefore, the same antigenic specificity is retained. Switching involves the deletion of intervening DNA between specific recombination sites called *switch regions.* Because the intervening DNA is lost, the B cell cannot "switch back" to an isotype that has already been deleted. The V region and C regions are transcribed together, and RNA splicing and translation results in expression of the *new* isotype.

The second process is termed **affinity maturation,** in which germline DNA is subject to mutational change, allowing coding for antibodies with increased affinity for binding to antigen. This process is a result of somatic hypermutation in which V regions of the antibody heavy and light chain genes undergo more than 10,000 times higher rate of mutation than "regular" DNA. Somatic hypermutation occurs only after antigen stimulation. Some of these mutations increase the affinity of antibody for antigen, and those B cells expressing antibody with higher affinity will be selectively stimulated, increasing the proportion of high-affinity antibody in secondary responses.

KEY POINTS ABOUT ANTIBODY DIVERSITY

- The V(D)J recombination that occurs during B-cell development, along with somatic mutation after antigenic stimulation, leads to the generation of antigen-binding diversity.

- Each individual B cell and all of its progeny express only one heavy chain and one light chain V region sequence; thus all have the same antigenic specificity.

Continued

- Plasma cells are activated B lymphoctyes that secrete antibody molecules.
- Isotype switching allows functional diversity due to unique properties associated with each antibody isotype.
- The sources of antibody diversity include the presence of multiple V gene segments, combinatorial diversity resulting from random recombination of V, D, and J segments, diversity due to insertion of nucleotides which result in amino acid changes in the V-D and D-J junctions, and the coexpression of different heavy and light chain pairs.
- Somatic hypermutation allows affinity maturation after the plasma cell has encountered antigen and undergone associated stimulation.

KEY POINTS

- The basic immunoglobulin (antibody) unit consists of two light and two heavy chains, all covalently bound via disulfide bridges. Domains of the molecule are responsible for unique biologic functions.
- There are five classes or isotypes of heavy chains—γ, α, μ, δ, and ϵ—corresponding to the heavy chains of IgG, IgA, IgM, IgD, and IgE, respectively. There are two types of light chains in humans—κ and λ—which are distinguished by different amino acid sequences in the constant domain.
- Antibodies recognize antigenic determinants (epitopes); the major classes of antigens include proteins or glycoproteins, nucleic acids, carbohydrates, and lipids.
- Antibody-binding to antigen occurs in hypervariable regions and does not use covalent interactions. Multiple weak forces stabilize the binding, including electrostatic interactions, hydrogen bonding, hydrophobic interactions, and van der Waals forces.
- Generation of antigen-binding diversity results from V(D)J recombination during B-cell development, and somatic mutation after antigenic stimulation. Biofunctional diversity of antibodies results from isotype switching, with unique biofunctions associated with each antibody isotype. Each individual B cell and all of its progeny express only one heavy chain and one light chain V region sequence; thus all have the same antigenic specificity.
- There are multiple sources of antibody diversity: (1) the presence of multiple V gene segments; (2) combinatorial diversity, resulting from random recombination of V, D, and J segment combinations; (3) junctional and insertional diversity, resulting in changes in the V-D and D-J junctions; (4) coexpression of different heavy and light chain pairs; and (5) somatic hypermutation.

Self-assessment questions can be accessed at www.StudentConsult.com.

T-Cell Immunity 4

CONTENTS
- ANTIGEN-SPECIFIC RECEPTORS
- GENES CODING FOR T-CELL RECEPTORS
- DEVELOPMENT OF T LYMPHOCYTES
- UNIQUE EFFECTOR FUNCTIONS OF T-CELL PHENOTYPES
  - T Helper Cells
  - T Cytotoxic Cells
  - T Regulatory/Suppressor Cells
  - Natural Killer T Cells
- γδ T CELLS

Thymus-dependent T lymphocytes are involved in the regulation of immune response and cell-mediated immunity. Overall, they regulate the major properties of the acquired immune response, leading to specificity of antigenic recognition and discrimination between self and nonself. There are two major classes of T cells; T helper T_H cells and cytotoxic T cells (CTLs). T_H cells are further divided into classes that assist in the activation of antigen-presenting cells (APCs) to kill intracellular bacteria or to help B cells produce antibody. CTLs primarily target virally infected cells and tumor cells.

ANTIGEN-SPECIFIC RECEPTORS

Mature T cells express T-cell receptors (TCRs) specific for antigen (Fig. 4-1). The TCR is expressed on the cell surface in association with coreceptors and accessory molecules that allow T cells to recognize presented antigens associated with the major histocompatibility complex (MHC) molecules on target cells. The TCR is a transmembrane heterodimer composed of two disulfide-linked polypeptide chains. T cells of all antigenic specificities exist prior to contact with antigen. Each lymphocyte carries a TCR of only a single specificity, which, when stimulated by antigen, gives rise to progeny with identical antigenic specificity.

The vast majority of peripheral T cells express alpha (α) and beta (β) chains on their surface. Cells that express gamma (γ) and delta (δ) chains make up only 5% of the normal circulating T-cell population in healthy adults. Each chain (α, β, γ, or δ) represents a distinct protein with an approximate molecular weight of 45 kDa. An individual T cell can express either an αβ or a γδ heterodimer as its receptor, but never both.

The TCR recognizes antigen in the form of short linear peptides that are bound in an antigen-binding pocket, or "groove," in the MHC molecules. The interactions between the TCR heterodimers create three hypervariable regions called complementarity-determining regions (1, 2, and 3), which confer specificity for antigen recognition (Fig. 4-2). The TCR heterodimer is in tight noncovalent association with the CD3 complex, which is made up of independently encoded subunits (one δ, γ, ε, and two ζ). The CD3 complex is required for efficient transport of the TCR to the cell surface. CD3 subunits possess long intracellular tails and are responsible for transducing signals upon TCR engagement of antigen in the context of MHC. Other coreceptors enhance the ability of T cells to respond to antigen; CD4 molecules assist helper cell interactions, while CD8 molecules contribute to CTL activity. They do this by binding to invariant regions of MHC class II and I molecules (see Chapter 5), respectively, on adjacent cells. Numerous molecules on the T-cell surface, including integrin cell adhesion molecules and the immunoglobulin family member CD28, promote T-cell adhesion to target cells or APCs (facilitating antigen recognition) and provide necessary costimulatory signals that further drive T-cell responses.

KEY POINTS ABOUT T LYMPHOCYTES

- T lymphocytes are involved in regulation of immune response and in cell-mediated immunity.
- Mature T cells express antigen-specific TCR in a complex with CD3 molecules. The TCR is a disulfide-linked heterodimer composed of either αβ or γδ chains. T cells express either αβ or γδ chain heterodimers, but never both.

BIOCHEMISTRY

Tyrosine Kinases

The cytoplasmic domains of CD3 molecules associated with the TCR are characterized by the presence of immunoreceptor tyrosine-based activation motifs. Signals delivered during antigen recognition trigger tyrosine kinases, which play important regulatory roles in cellular activation, leading to activation of other downstream effectors such as phospholipase C and proteins that activate mitogen-activated protein kinases.

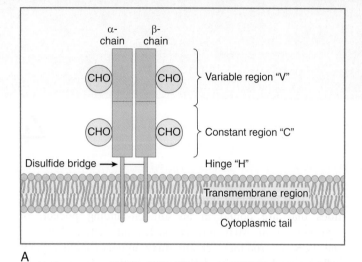

Figure 4-1. Structure of the T-cell receptor (*TCR*) complex. **A,** The structure of the TCR complex showing the predominant form of the antigen-binding chains, α (40–50 kDa) and β (35–47 kDa). A T cell expresses a TCR of only one specificity. Both transmembrane peptides exhibit two external domains connected via a disulfide bridging bond. A hydrophobic helical domain crosses the cellular membrane, leaving a short cytoplasmic tail. **B,** The TCR is always expressed with the associated CD3 complex, required for signal transduction, composed of γ, δ, and ε chains plus two ζ chains. T helper cells express CD4, required for interaction with antigen-presenting cells, whereas cytotoxic T cells express the CD8 coreceptor molecule. *CHO*, carbohydrate.

●●● GENES CODING FOR T-CELL RECEPTORS

Genes that code for TCRs and the mechanisms used to generate TCR diversity are similar to those of immunoglobulins. In fact, the TCR genes are closely related members of the immunoglobulin gene superfamily. Each chain consists of a constant (C) and a variable (V) region. The repertoire is generated by combinatorial joining of variable (V), joining (J), and diversity (D) genes and by N region diversification (nucleotides inserted by the enzyme deoxynucleotidyl transferase)

(Fig. 4-3). Unlike immunoglobulin genes, genes encoding TCR do not undergo somatic mutation. Thus, there is no change in the affinity of the TCR during activation, differentiation, and expansion.

The TCR V, D, and J genes are mixed together in a more complicated manner than is found for immunoglobulin genes. For example, the α and γ chains use only V and J gene segments, whereas the β and δ chains use V, D, and J gene segments. There are many more Vα and Vβ genes (50 to 100) than Vγ and Vδ genes (5 to 10) present in germline. The α and δ chain genes are mixed together in one locus. The genes encoding the δ chain are located entirely between the cluster of Vα and Jα gene segments.

Recombination of V, D, and J gene segments is coordinated by recombinase-activating genes *RAG-1* and *RAG-2*. The enzymes recognize specific DNA signal sequences consisting of a heptamer, followed by a spacer of 12 or 23 bases, and then by a nonamer. If either *RAG* gene is impaired, mutated, or missing, homologous recombination events are abolished. This gives rise to severe combined immunodeficiency. An example of such a disorder is Omenn syndrome, an autosomal recessive form of severe combined immunodeficiency characterized by high susceptibility to fungal, bacterial, and viral infections.

The order of gene rearrangement is sequential. An immature T lymphocyte entering the thymus has its TCR genes in the germline configuration because rearrangement has not yet taken place. Both γ and β chain genes then begin to rearrange, more or less simultaneously. If the γ chain genes rearrange successfully, then δ chain genes also start to rearrange. If both γ and δ genes rearrange functionally, no further gene rearrangement takes place and the cell remains a γδ T cell. If γ and/or δ rearrangements are not functional, then β gene rearrangement continues followed by α gene rearrangement. In this manner, an αβ product appears and the cell becomes an αβ T cell.

The overall level of diversity is greater for TCRs than for immunoglobulins; up to 10^{18} possible antigen-specific TCRs are generated compared with 10^{14} possible antibody specificities (Fig. 4-4). The increased T-cell diversity is due to additional junctional regions and greater N nucleotide addition during recombination, coding for regions corresponding to the CDR3 loops that form the center of antigen-binding sites. Although there is great variability in the binding site regions, the remaining framework portion of the heterodimer is subject to relatively little variation.

KEY POINTS ABOUT TCR GENES

- TCR genes are closely related members of the immunoglobulin gene superfamily and derive part of their structural diversity from recombination of different V, D, and J gene segments.
- The process of recombination is coordinated by recombinase-activating genes *RAG-1* and *RAG-2*.
- Unlike the B-cell antibody genes, TCR genes do not undergo somatic mutations.

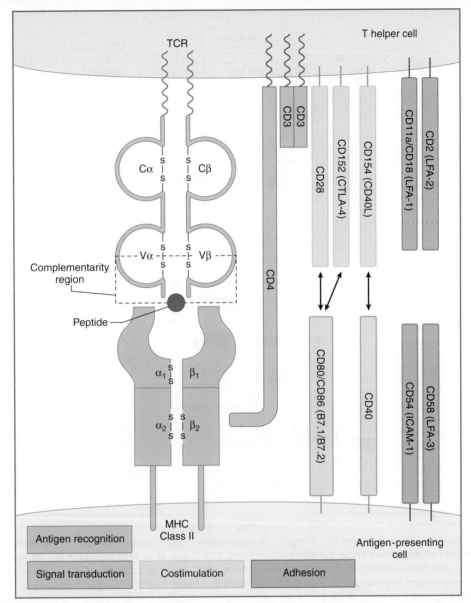

Figure 4-2. Interaction of T-cell receptor (*TCR*), major histocompatability complex (*MHC*), and peptide. The complementarity-determining regions of the TCR variable regions and peptide bound in the binding groove of an MHC class II molecule are depicted. The TCR recognizes portions of both the MHC molecule and the bound antigen. For T helper cells, recognition of the presented antigen is enhanced by CD4, which attaches to nonpolymorphic regions on the MHC molecule. Further stabilization is accomplished by integrin interactions between molecules on the T cell with ligands on the surface of the antigen-presenting cell. Costimulatory molecules present on both the T cell and the antigen-presenting cell are critical for T-cell function and activation.

PATHOLOGY

Severe Combined Immunodeficiency

Lymph node biopsies in severe combined immunodeficiency patients with Omenn syndrome demonstrate an absence of germinal centers and replacement of the paracortex by a diffuse infiltrate of large cells with abundant pale cytoplasm.

DEVELOPMENT OF T LYMPHOCYTES

Progenitor cells from the bone marrow migrate to the thymus, where they differentiate into T lymphocytes (Fig. 4-5). In the thymus, immature T cells undergo rearrangement of their TCR genes to generate a diverse set of clonotypic TCRs. Immature thymocytes are selected for further maturation only if their TCRs do not interact with self-peptides presented

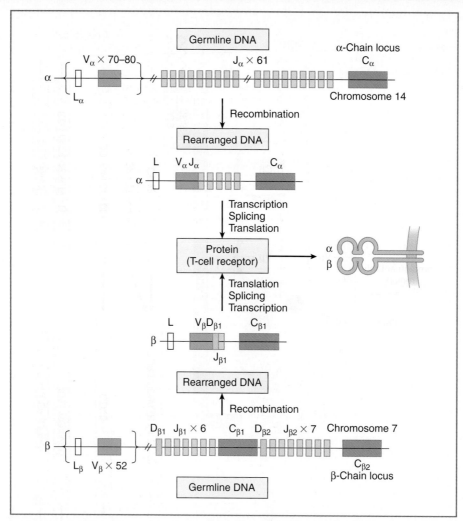

Figure 4-3. Gene rearrangement of T-cell receptor (*TCR*) genes. TCR diversity is generated by combinatorial joining of variable (*V*), joining (*J*), and diversity (*D*) genes and by N region diversification (nucleotides inserted by the enzyme deoxynucleotidyl transferase). The top and bottom rows show germline arrangement of the V, D, J, and constant (*C*) gene segments at the TCR α and β loci. During T-cell development, a V region sequence for each chain is assembled by DNA recombination. For the α chain (*top*), a Vα gene segment rearranges to a Jα gene segment to create a functional gene encoding the V domain. For the β chain (*bottom*), rearrangement of a Dβ, a Jβ, and a Vβ gene segment creates the functional V domain exon.

in the context of self-MHC molecules on APCs. Positively selected thymocytes undergo commitment to either the T cytotoxic or T_H lineages. Deletion of self-reactive T cells occurs through apoptosis. Lineage commitment is marked phenotypically by the loss of expression of one of the coreceptor molecules, CD8 or CD4. Immature thymocytes express both coreceptors (double-positive), whereas T killer or T_H cells express only CD8 or CD4, respectively (single-positive CD8+ or CD4+). T lymphocytes continue to differentiate after leaving the thymus, driven by encounter with specific antigen in the secondary lymphoid organs (spleen, lymph nodes, and mucosa-associated lymphoid tissue).

Virtually all the cells that express the TCR-αβ are CD4+ (T_H) or CD8+ (T cytotoxic or T suppressor). Almost all cells expressing TCR-γδ are CD4− CD8− (double-negative). The TCR-αβ-expressing lymphocytes are known to function as effector cells, whereas the function of the TCR-γδ cells is not well understood.

PHYSIOLOGY

Apoptosis

Apoptosis, or programmed cell death, is the regulated elimination of cells that occurs naturally during the course of development, as well as in many pathologic circumstances that require cell death for the benefit of the organism. For example, during vertebrate embryonic development, the notochord and the floor plate secrete a gradient of the signaling molecule Sonic Hedgehog (Shh), which directs cells to form patterns in the embryonic neural tube.

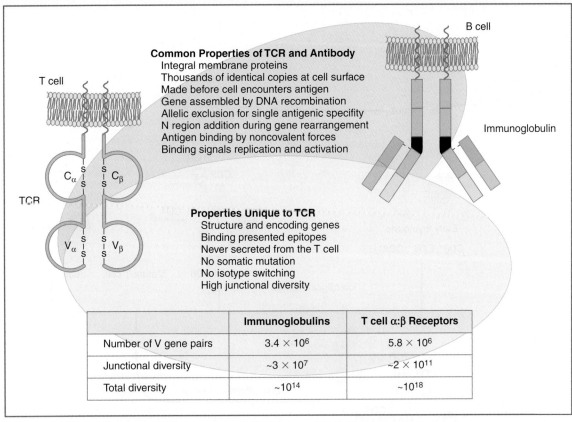

Figure 4-4. Comparison of diversity between B- and T-cell receptors. *TCR,* T-cell receptor.

●●● UNIQUE EFFECTOR FUNCTIONS OF T-CELL PHENOTYPES

T Helper Cells

T_H cells are the primary regulators of T-cell– and B-cell–mediated responses (Fig. 4-6). They (1) aid antigen-stimulated subsets of B lymphocytes to proliferate and differentiate toward antibody-producing cells; (2) express the CD4 coreceptor molecule; (3) recognize foreign antigen complexed with MHC class II molecules on B cells, macrophages, or other APCs; and (4) aid effector T lymphocytes in cell-mediated immunity. Currently, it is believed that there are two main functional subsets of T_H cells, plus other helper subsets of increasing importance (Table 4.1). As a general rule, T_H1 cells aid in the regulation of cellular immunity, and T_H2 cells aid B cells to produce certain classes of antibodies (e.g., IgA and IgE) and release cytokines that target mast cells and eosinophils. The functions of these subsets of T_H cells depend upon the specific types of cytokines that are generated, such as interleukin-2 (IL-2) and interferon-γ by T_H1 cells IL-4, IL-6 and IL-13 by T_H2 cells. Two other classes of T_H cells, T_H17 and T regulatory cells (**Tregs cells**), are thought to serve as regulators for immune function. The T_H17 phenotypic population is characterized by IL-17 secretion; these cells are thought to be involved as effector cells for autoimmune disease progression and as well as important in the enhancement of host protection against extracellular pathogens. The Treg population (discussed below) serves to regulate overall T-cell responsiveness. Other subclasses, such as T_H3, Tr1, and T_H9 phenotypes, have been reported but probably represent diversity within other T-cell lineages. Of distinction, the T_H3 cells secrete IL-10 and tumor growth factor—β, provide help for IgA production, and regulate mucosal immune function.

KEY POINTS ABOUT T_H LYMPHOCYTES

- T_H lymphocytes recognize specific antigens in the context of MHC class II on APCs. Although this is required for T-cell activation, it is not sufficient. Other accessory molecules on T cells, such as integrins and CD28, which bind counter-ligands on the adjacent cell, promote firm adhesion between the T cell and APC and send necessary secondary signals that fully activate T cells.
- Activation of the T_H cell leads to specific cytokine release; different classes of T_H cells secrete specific subsets of cytokines.

Cytotoxic T Cells

CTLs are cytotoxic against tumor cells and host cells infected with intracellular pathogens (Fig. 4-7). These cells (1) express the CD8 coreceptor and (2) destroy infected

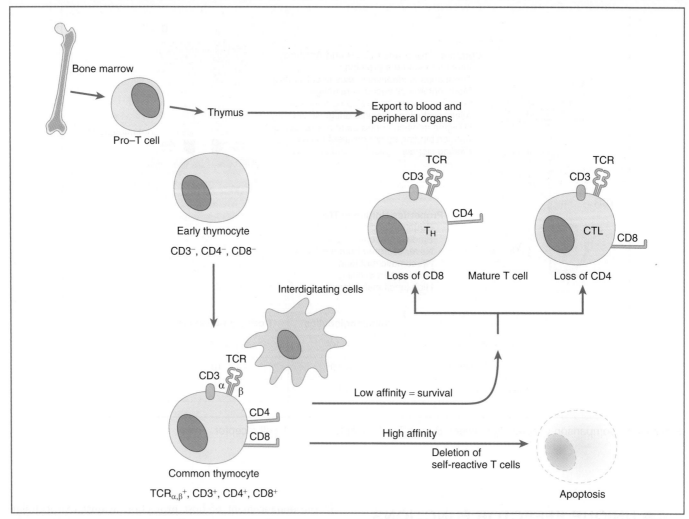

Figure 4-5. Main stages in thymic selection. Pro-T cells migrating from the bone marrow enter the thymus, where they begin to express rearranged T-cell receptor (*TCR*), and CD3, CD4, and CD8 proteins. Positive selection occurs to eliminate self-reactive T cells through mechanisms of apoptosis. Maturing thymocytes lose either CD4 or CD8 surface molecules, and they are exported to peripheral tissue as CD4+ T helper cells or as cytotoxic T cells.

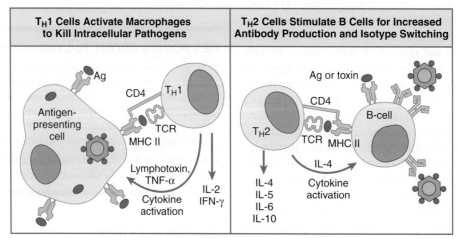

Figure 4-6. Effector functions of T helper (T_H) cells. T_H1 cells activate macrophages through production of cytokines, leading to assisted destruction of intracellular microorganisms. T_H2 cells drive B-cell differentiation to stimulate B cells to proliferate, to secrete immunoglobulins, and to undergo isotype switching events. *Ag,* antigen; *IFN,* interferon; *IL,* interleukin; *MHC,* major histocompatability complex; *TCR,* T-cell receptor; *TNF-α,* tumor necrosis factor-α.

TABLE 4-1. Characteristics of Two of the Major T Helper Cell Subsets

T CELL POPULATION	MAJOR CYTOKINE PROFILE	EFFECTOR FUNCTION	FACTORS FOR GROWTH AND DEVELOPMENT	NEGATIVE REGULATOR
T_H1	IL-2, IL-3, IL-15, IFN-γ, TNF-α, TNF-β, GM-CSF	Cellular immunity, NK cell activation	IL-2, IL-12, IL-18, IL-27, TNF-α	IL-10, TGF-β
T_H2	IL-4, IL-5, IL-6, IL-10, IL-13, IL-21, TNF-β	B-cell maturation	IL-4	IFN-γ

T_H, T helper; *IL*, interleukin; *INF*, interferon; *TNF*, tumor necrosis factor; *GM-CSF*, granulocyte colony-stimulating factor; *NK*, natural killer.

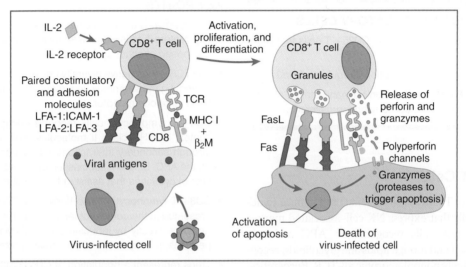

Figure 4-7. Effector function of cytotoxic T cells (CTLs). CTLs recognize virally infected target cells, which express foreign antigens complexed with major histocompatability complex (*MHC*) class I molecules. Cytotoxic effector molecules produced by the CTL (perforins, granzymes) initiate destruction of the target cell and deliver apoptotic signals through Fas and FasL on cellular surfaces. β_2M, β_2-microglobulin; *LFA*, lymphocyte function-associated antigen; *ICAM*, intracellular adhesion molecule; *TCR*, T-cell receptor.

cells in an antigen-specific manner that depends on the expression of MHC class I molecules on APCs. CTLs are able to kill target cells directly by inducing apoptosis. Nucleases and other enzymes activated in the apoptotic process may help destroy the viral genome, thus preventing the assembly of virions and potential infection of other cells. CTLs induce apoptosis only in the target cell; neighboring tissue cells are not affected. Two mechanisms for induction of apoptosis have been identified. One method uses preformed perforins directionally released from the CTL to the target cell surface that generate transmembrane pores, through which a second group of proteins, granzymes, can gain entry to the cytosol and induce an apoptotic series of events. The other method occurs by apoptotic signaling via membrane-bound Fas molecules on the target cell surface and Fas ligand on the CTL surface. The processes of antigen recognition, CTL activation, and delivery of apoptotic signals to the target cell can be accomplished within 10 minutes. The apoptotic process in the targeted cell may take 4 hours or more and continues long after the CTL has moved on to interact with other potential targets.

KEY POINTS ABOUT CYTOTOXIC T CELLS

- CTLs recognize antigen in the context of MHC class I and, like T_H cells, are fully activated by accessory costimulatory molecules.
- CTLs kill target cells directly by inducing apoptosis. They release preformed perforins at the target cell surface to generate transmembrane pores in the target cell, through which a second set of proteins and granzymes gain entry into the cytosol to initiate an apoptotic series of events.
- CTLs can also deliver apoptotic signals via surface-bound molecules.

T Regulatory/Suppressor Cells

Historically, cytotoxic suppressor cells were identified with function in regulating cellular (T cell) responses. Recently, it has become clear that this group represents a separate subpopulation of CD4+ lymphocytes generated in the thymus, which reflects consequent activities of T cells to produce cytokines that downregulate developmental pathways of T_H responsiveness. This population is identified by expression of both the CD4 marker and the IL-2 receptor α chain (CD25), often with co-expression of CD45. Specific engagement of the TCR is required for function. Regulation is controlled via the transcriptional regulator *Foxp3*, mutations in which result in incidence of autoimmunity as well as uncontrolled lymphoproliferation. Mechanisms of immunosuppression/tolerance by CD4(+)CD25(+) Tregs include the local secretion of cytokines such as tumor growth factor–β and IL-10, and direct cell contact through binding of cell surface molecules such as CTLA-4 (CD152) on Tregs to CD80 and CD86 molecules on effector T cells.

KEY POINTS ABOUT T REGULATORY CELLS

- Tregs represent a population of CD4+CD25+ T cells that regulate effector T cells.
- Tregs function through the transcriptional regulator *Foxp3*.
- Loss or mutation in *Foxp3* results in clinical manifestation of autoimmune disease and/or uncontrolled lymphopro-liferation.

Natural Killer T Cells

The term natural killer T (NKT) cell was first used in mice to define a subset of T cells that express NK cell–associated marker NK1.1 (CD161). These cells recognize an APC molecule (CD1) that presents self and foreign lipids and glycolipids, recognized through a relatively nonpolymorphic αβ-TCR. Recognition of these antigens does not require CD4 or CD8 coreceptors for activation. NKT cells are able to produce large quantities of INF-γ or IL-4 as well as multiple other cytokines and chemokines. Dysfunction or deficiency in NKT cells has been linked to the development of autoimmune diseases (such as diabetes or atherosclerosis), progression to asthma, and development of certain cancers. Of interest, a potent agonist for cell activation of NKT cells is α-galactosylceramide, which is now being examined for broad-based use in development as a vaccine adjuvant.

γδ T CELLS

γδ T cells are predominantly found in the lamina propria of the gut and are thought to assist in protection against microorganisms entering through epithelium at mucosal surfaces. Their range of response to antigens is limited. These cells have been found to be active toward mycobacterial antigens and heat shock proteins, and they have the ability to secrete cytokines like their αβ counterparts. Generally, γδ cells lack CD4 and CD8 (double-negative cells), although some γδ cells do express CD8. Some γδ T cells can function in the absence of MHC molecules.

KEY POINTS

- T lymphocytes are the master regulators of cell-mediated immunity. Mature T cells express an antigen-specific TCR heterodimer composed of either αβ or γδ chains.
- The TCR genes derive diversity from recombination of different V, D, and J gene segments, in a manner similar to that seen for the B-cell antibody receptor.
- CD4+ T lymphocyte cells recognize antigen presented in the context of MHC class II molecules. Full activation is obtained through secondary interactions with surface molecules present on the APC. Different subclasses of T_H cells secrete unique subsets of cytokines that assist in functional responses.
- CD8+ T lymphocytes, also known as CTLs, recognize antigen presented in the context of MHC class I molecules. CTLs kill target cells directly by inducing apoptosis via released preformed perforins, granzymes, and secondary proteins.
- Tregs represent a population of CD4+CD25+ T cells that regulate effector T cells through action of the transcriptional regulator *Foxp3*.
- NKT cells are a subset of T cells that express NK cell markers. NKT cells recognize lipids and glycolipids using a relatively non-polymorphic αβ-T cell receptor.

Self-assessment questions can be accessed at www.StudentConsult.com.

Role of Major Histocompatibility Complex in the Immune Response

5

CONTENTS

FUNCTION OF THE MAJOR HISTOCOMPATIBILITY COMPLEX

ORGANIZATION AND HIGH POLYMORPHISM OF MHC GENES AND GENE PRODUCTS

STRUCTURE OF MHC CLASS I AND CLASS II MOLECULES

MHC AND ANTIGEN PRESENTATION
  Endogenous (Cytoplasmic) Antigen Processing and MHC Class I Presentation
  Exogenous (Endosomal) Antigen Processing and MHC Class II Presentation

ROLE OF MHC IN THYMIC EDUCATION

ROLE OF MHC IN ACTIVATION OF T CELLS

ASSOCIATION OF DISEASE WITH MHC HAPLOTYPE

ATYPICAL AND UNIQUE ANTIGEN PRESENTATION
  Superantigens Directly Bind T-Cell Receptors and MHC Without Processing
  Presentation of Nonclassical Lipid Antigens to T Cells

●●● FUNCTION OF THE MAJOR HISTOCOMPATIBILITY COMPLEX

The major histocompatibility complex (MHC) is a gene locus composed of multiple sequences that code for histocompatibility antigens. Histocompatibility antigens are cell surface glycoproteins that play critical roles in interactions among immune system cells. MHC genes are organized into two major classes. Class I gene products are found on all nucleated cells and participate in antigen presentation to CD8+ cytotoxic lymphocytes (CTL). Class II molecules are found primarily on antigen-presenting cells (APCs; B cells, macrophages, and dendritic cells) and participate in antigen presentation by macrophages to CD4+ lymphocytes (T helper cells). MHC genes are highly polymorphic. In humans, the MHC locus is designated as the human leukocyte antigen (HLA) locus.

MHC molecules were so named because they were first identified as the targets for rejection of grafts between individuals. When organs are transplanted between a donor and a recipient with MHC locus differences, graft rejection is prompt. The function of the MHC is the presentation of antigen fragments (epitopes) to T cells.

●●● ORGANIZATION AND HIGH POLYMORPHISM OF MHC GENES AND GENE PRODUCTS

The HLA locus in humans is found on the short arm of chromosome 6 (Fig. 5-1). The class I region consists of HLA-A, HLA-B, and HLA-C loci. The class II region consists of the D region, which is subdivided into HLA-DP, HLA-DQ, and HLA-DR subregions. A region between the class I and class II loci encodes for class II proteins with no structural similarity to either class I or class II molecules (complement proteins, tumor necrosis factor, and lymphotoxin).

The highly polymorphic class I and class II MHC products are central to the ability of T cells to recognize foreign antigen and the ability to discriminate "self" from "nonself." MHC class I and class II molecules that are not possessed by an individual are seen as foreign antigens upon transplantation and are dealt with accordingly by the recipient's immune system. All MHC molecules show a high level of allotypic polymorphism; that is, certain regions of the molecules differ from one person to another. The chance of two unrelated people having the same allotypes at all genes that encode MHC molecules is very small.

The class I MHC molecules are each somewhat different from one another with respect to amino acid sequence, and all three (HLA-A, HLA-B, and HLA-C) are codominantly expressed in the membrane of every nucleated cell (as high as 5×10^5 molecules per cell on lymphocytes). *Codominantly expressed* means that each gene encoding these proteins, from both parental chromosomes, is expressed. MHC class I

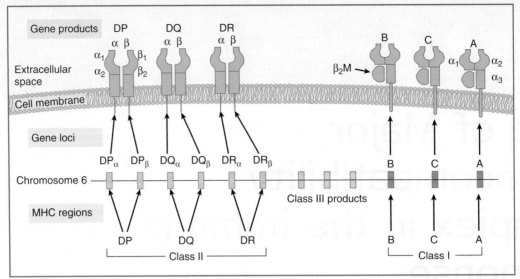

Figure 5-1. Genetic organization of the human leukocyte antigen (HLA) locus and associated gene products. The polymorphic human major histocompatability complex (*MHC*) genes of the HLA locus coding for class I and class II molecules are located on chromosome 6. Class I genes are designated as A, B, and C, each coding for a three-domain polypeptide (α_1, α_2, α_3) that associates with invariant β_2-microglobulin. Class II genes are DP, DQ, and DR, each coding for individual α and β chains that interact to provide binding sites for antigen presentation. $\beta_2 M$, β_2-microglobulin.

molecules expressed on progeny (F1) cells match maternal or paternal class I molecules. The MHC class II molecules expressed on F1 cells include homologous and heterologous $\alpha\beta$ dimer mixtures, since both α and β subunit genes exhibit species-specific polymorphism. Homologous dimers match class II molecules expressed on either parental cell type, whereas heterologous dimers are unique to the F1 genotype and are functionally nonequivalent to parental class II molecules.

KEY POINTS ABOUT MHC GENES AND GENE PRODUCTS

- MHC molecules are important for antigen presentation to T cells.
- Human class I genes consist of the HLA-A, HLA-B, and HLA-C loci; associated gene products are found on all nucleated cells.
- The class II region genes consist of the D region, which is subdivided into HLA-DP, HLA-DQ, and HLA-DR subregions; the class II molecules are found on APCs.
- All MHC molecules show high allotypic polymorphism.

GENETICS

Codominant Gene Expression

Codominant expression of genes defined for pairs of heterozygous alleles leads to the situation in which both phenotypes may be present, or the expression of two alleles results in a blend of traits. For example, the codominantly expressed blood type AB is qualitatively distinct from either parental type A or B.

STRUCTURE OF MHC CLASS I AND CLASS II MOLECULES

Each class I locus codes for a transmembrane polypeptide of molecular weight approximately 45 kDa, containing three extracellular domains (α_1, α_2, α_3). The molecule is expressed at the cell surface in a noncovalent association with an invariant polypeptide called β_2-microglobulin ($\beta_2 M$) of 12 kDa. $\beta_2 M$ is a member of the Ig superfamily; the complex of class I and $\beta_2 M$ is expressed as a molecule with four domains associated such that the $\beta_2 M$ and α_3 domain of class I are juxtaposed near the cell surface membrane (Fig. 5-2).

Class II molecules have two transmembrane polypeptide chains (α and β, 30 to 34 and 26 to 29 kDa, respectively); the peptide-binding site is shared by the two domains farthest from the cell membrane. The overall structure of the peptide-binding site is very similar for both class I and class II MHC molecules; the base is made of β-pleated sheet, as in an immunoglobulin domain; the sides of the groove that holds the peptide are α-helices. Peptides bind within the allele-specific pockets defined by the two transmembrane polypeptide chains, where they are presented to the T-cell receptor (TCR) for recognition. The extracellular domain shows variability in amino acid sequences, yielding grooves with different shapes. These grooves cradle the processed antigen for interaction with the TCR (Fig. 5-3). The class II MHC molecule is capable of presenting peptides of a slightly larger size than that presented by the class I MHC complex.

Class I and class II genes also exhibit polymorphism with multiple allelic forms expressed. In humans, allelic forms are designated in a descriptive manner. For example, human class II genes are given numbers, such as HLA-D4 or HLA-D7, depending on loci location.

Structure of MHC class I and class II molecules

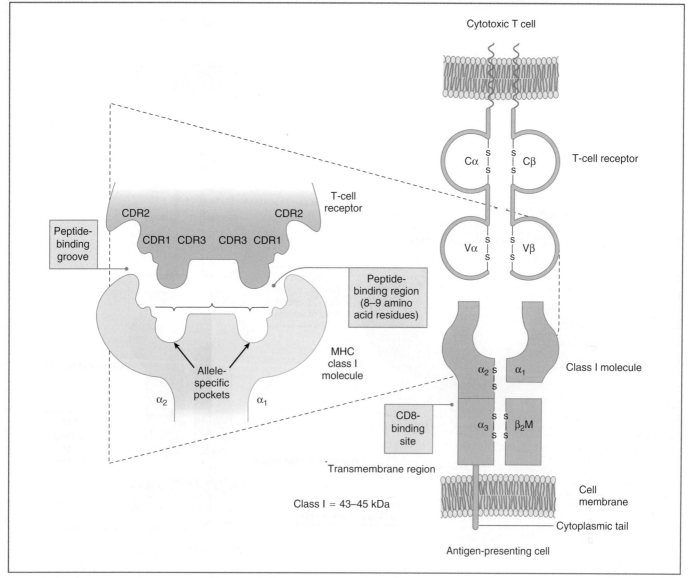

Figure 5-2. Major histocompatability complex (*MHC*) class I. The class I MHC transmembrane molecule is associated with the invariant β₂-microglobulin, giving structure to the extracellular domain for presentation of processed antigen to cytotoxic T cells. Interaction of the T-cell receptor is stabilized by CD8 recognition of the class I molecule. Shown here is a short peptide fragment (8 to 10 amino acids in length) noncovalently interacting with the α_1 and α_2 domains on the class I molecule and complementarity-determining regions on the T-cell receptor. *β₂M*, β₂-microglobulin.

KEY POINTS ABOUT STRUCTURE OF MHC CLASS I AND CLASS II MOLECULES

- Class I molecules are important in presentation of antigen epitopes to CD8+ T cells. Each class I locus codes for a transmembrane polypeptide containing three extracellular domains (α_1, α_2, α_3), which is expressed at the cell surface in a noncovalent association with β₂M.
- Class II molecules present antigen to CD4+ lymphocytes. Class II molecules have two transmembrane polypeptide chains. Peptides bind within the allele-specific pockets defined by the two transmembrane polypeptide chains, where they are presented to the TCR for recognition.

MHC and Antigen Presentation

Antigen is recognized by T cells when it is presented in complex with proteins of the MHC. Different antigen degradation and processing pathways produce MHC-peptide complexes in which endogenous (cytoplasmic) peptides associate with class I molecules and exogenous (endosomal) peptides associate with class II molecules. MHC class I presents endogenous antigen epitopes to CD8+ T cytotoxic cells, and MHC class II presents exogenous antigen epitopes to CD4+ T_H cells. All nucleated cells are capable of presenting MHC class I, but only specialized phagocytic cells present antigenic epitopes on MHC class II.

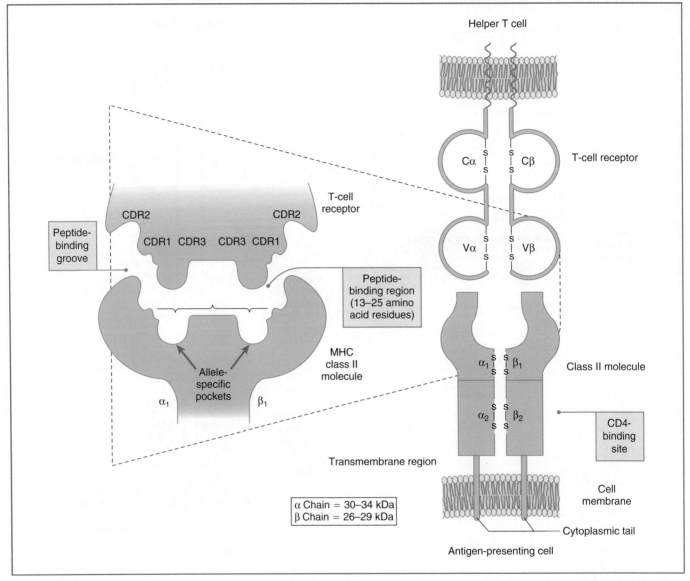

Figure 5-3. Major histocompatability complex (*MHC*) class II. Processed antigenic fragments (13 to 25 amino acids in length) interact with the α_1 and β_1 domains on the class II molecule within the peptide-binding groove, allowing presentation to CD4+ T helper cells. Interactions with the T-cell receptor are stabilized by CD4 recognition of conserved regions on the class II molecule.

Endogenous (Cytoplasmic) Antigen Processing and MHC Class I Presentation

MHC class I molecules bind peptide fragments derived from proteolytically degraded proteins endogenously synthesized by a cell (Fig. 5-4). Examples of endogenously produced antigens include proteins produced during intracellular viral replication. Small peptides are transported into the endoplasmic reticulum, where they associate with nascent MHC class I molecules before being routed through the Golgi apparatus and displayed on the surface for recognition by cytotoxic T lymphocytes. MHC class I molecules bind small antigenic peptides that are 8 to 10 amino acid residues in length. Upon recognition by CTLs, the CD8 molecule binds to constant domains on the class I α chain to stabilize the interaction.

When a T-cell (CTL) receptor binds antigen-MHC class I and costimulatory signals occur, the CTL is activated to produce and secrete toxin that kills the cell to which it is bound.

BIOCHEMISTRY

Adenosine Triphosphate–Mediated Intracellular Transporters

The transporters associated with antigen presentation are cytosolic proteases that shuttle peptides into the exocytic compartment for association with nascent MHC class I molecules. Specifically, TAP-1 and TAP-2 are members of the adenosine triphosphate binding cassette transporter superfamily, which uses adenosine triphosphate to drive the transport of molecules across all cell membranes.

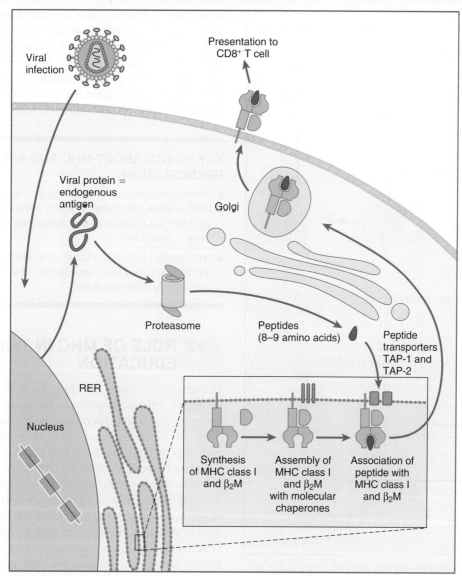

Figure 5-4. Endogenous pathway for antigen processing and class I presentation. Class I major histocompatability complex (*MHC*) molecules interact with peptides degraded by proteasomes, part of a large cytoplasmic proteolytic complex called the *low-molecular-mass polypeptide*. Degraded peptides are carried into the rough endoplasmic reticulum by transporters of antigenic peptides (*TAP-1, TAP-2*). Upon peptide binding, the interaction between a class I MHC α chain and β2-microglobulin (*β2M*) is stabilized and the complex is routed through the Golgi apparatus to the plasma membrane.

Exogenous (Endosomal) Antigen Processing and MHC Class II Presentation

MHC class II molecules bind peptide fragments derived from proteolytically degraded proteins exogenously internalized by APCs, including macrophages, dendritic cells, and B cells (Fig. 5-5). Examples of exogenous antigens include degraded products resulting from phagocytosed bacterial agents. The resulting peptide fragments are compartmentalized in the endosome, where they associate with MHC class II molecules before being routed to the cell surface for recognition by helper T lymphocytes. MHC class II molecules bind larger antigenic peptides, usually 13 to 25 amino acid residues in length (or longer).

As with class I molecules, class II MHC molecules are synthesized in the rough endoplasmic reticulum. The class II α and β chains reside there as a complex with an additional polypeptide called the *invariant chain*. The invariant chain blocks the groove of the class II molecule and prevents endogenous antigens from binding there. The MHC–invariant chain complex is transported to an acidic endosomal or lysosomal compartment that contains a degraded antigen peptide. The invariant chain comes off the complex, exposes the groove of the class II molecule, and allows the antigen peptide to slip into the groove. The class II antigen-peptide complex is then transported to the surface of the APC, where it is available for interaction with CD4+ T_H cells. When the CD4+ TCR binds antigen-MHC class II, and when appropriate costimulatory molecules are engaged, the CD4+ T cell is

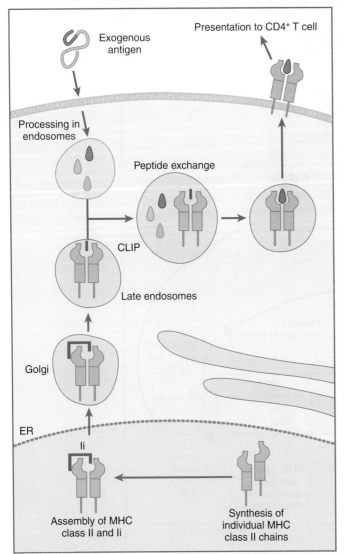

Figure 5-5. Exogenous pathway for antigen processing and class II presentation. The exogenous (endosomal) pathway prepares processed antigens for presentation to T cells via class II–regulated mechanisms. Nascent class II α and β polypeptides produced in the endoplasmic reticulum (*ER*) combine and interact with a specialized invariant chain that prevents peptide binding within this intracellular compartment. The invariant chain fills the peptide-binding pocket and facilitates routing of the complex to the endosomal compartment. Late in the exogenous processing pathway, endosomes containing class II major histocompatability complex (*MHC*) molecules and the invariant chain fuse with lysosomes. Enzymes within the lysosomes degrade the invariant chain, enabling the class II MHC molecules to bind peptides.

activated to proliferate and secrete cytokines, which in turn activate other cells of the immune system to generate humoral or cellular immunity.

B cells are excellent exogenous peptide presenter cells. The B cell can specifically phagocytose antigen via binding through surface immunoglobulin. Antigen is internalized, broken down to peptides presented on the B-cell surface in the peptide-binding grooves of MHC class II molecules. The subsequent interaction of a B cell with a primed T cell that recognizes the peptide–class II complex results in proliferative and activation pathways, which drive isotype switching as well as maturation of the B lymphocyte to become an antibody-secreting plasma cell. Importantly, since all nucleated cells express class I molecules, B cells have the innate ability to process antigen for incorporation of peptides into class I or class II presentation pathways.

KEY POINTS ABOUT MHC AND ANTIGEN PRESENTATION

- Different antigen degradation and processing pathways produce MHC-peptide complexes in which *endogenous* peptides associate with class I molecules and *exogenous* peptides associate with class II molecules.
- MHC class I molecules bind small antigenic peptides that are 8 to 10 amino acid residues in length; MHC class II molecules present slightly larger peptides.

ROLE OF MHC IN THYMIC EDUCATION

The education process by which T cells in the thymus learn to recognize antigenic peptides in the context of self-MHC molecules is a two-step procedure involving both positive and negative selection. Immature T cells (thymocytes) interact with thymic epithelial cells that express high levels of both class I and class II MHC molecules. Thymocytes with moderate affinities for these self-MHC molecules are allowed to develop further, whereas thymocytes with affinities too high or too low for self-MHC are induced to die by apoptosis. The thymocytes that survive are said to have been **positively selected** through their weak interaction with self-MHC. The positively selected thymocytes then begin to express high levels of TCR. A negative selection process occurs as the thymocytes travel deeper into the cortex, at the corticomedullary junction, and in the medulla of the thymus. At this point, self-antigens fill the MHC-peptide grooves. Only thymocytes that fail to recognize self-antigens are allowed to survive and proceed further along the maturation process, with the remainder undergoing apoptosis. This ensures elimination of self-reactive T cells. Eventually, T cells that survive the negative selection process lose either CD4 or CD8, becoming **single-positive** cells. Fewer than 5% of thymocytes survive selection and leave the thymus to take up residence in the secondary lymphoid organs.

KEY POINT ABOUT THE ROLE OF MHC IN THYMIC EDUCATION

- T cells in the thymus learn to recognize "foreign" antigenic peptides in the context of self-MHC molecules by a two-step process involving both positive and negative selection.

ROLE OF MHC IN ACTIVATION OF T CELLS

Binding between the TCR and the MHC–antigen peptide complex is highly specific and acts as the first signal to induce T-cell activation. Mature T cells do not respond to either self-MHC alone or to free peptide. Activated T cells differ from resting T cells in that they proliferate and secrete lymphokines or lytic substances. The affinity of the TCR for the MHC-antigen complex often is too low to fully activate the T cell; in addition to interactions with CD4 or CD8, numerous accessory molecules increase avidity between the T cell and APC by performing both adhesive and costimulatory signaling functions (Fig. 5-6).

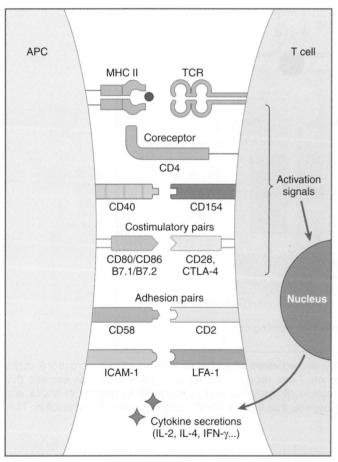

Figure 5-6. Accessory molecules involved in antigen presentation. The binding between the major histocompatability complex (*MHC*)–antigen peptide complex and the T-cell receptor (*TCR*) acts as the first signal toward induction of T-cell activation. Accessory molecules increase avidity between T cell and antigen-presenting cell (*APC*) by performing additional adhesive and signaling functions. Adhesion molecules, such as leukocyte function-associated antigen 1 (CD11a/CD18) and intracellular adhesion molecule-1 (*ICAM*) (CD54) pairing, promote firm adhesion between T cell and APC and can provide costimulatory signals for T-cell activation. This interaction also facilitates T-cell sampling of antigen in the context of MHC molecules on the surface of the APC. In addition, T cells require further costimulation through binding of CD28 with either of its ligands, CD80 or CD86 (B7.1 or B7.2). If all events transpire, full activation of the T cell occurs, leading to production of T-cell cytokines and a proliferative response. *IL*, interleukin; *IFN*, interferon.

HISTOLOGY

The Thymus

The cortex of the thymus has a "starry sky" appearance because of the presence of thymocytes and macrophages, whereas the medulla is characterized by the presence of Hassell's corpuscles (a key histologic characteristic of the thymus).

ASSOCIATION OF DISEASE WITH MHC HAPLOTYPE

Particular MHC alleles are associated with better protection against certain infections, while others have been identified as associated with a greater chance of developing autoimmunity. Some diseases are distinctly more common in individuals with a particular MHC allele or MHC haplotype. Diseases with a strong association with certain MHC alleles include type 1 diabetes and Graves disease. As another example, expression of HLA-DR4 is associated with rheumatoid arthritis. Nearly 90% of people with ankylosing spondylitis carry the HLA-B27 allele. Expression of HLA-DR2 is associated with multiple sclerosis. It is hypothesized that in some cases MHC molecules serve as receptors for the attachment and entry of pathogens into the cell; this makes individuals with a certain HLA type more susceptible to infection by a particular intracellular pathogen using that HLA molecule as a receptor. Alternatively, an infectious agent might possess antigenic determinants that resemble MHC molecules (molecular mimicry). Such resemblance might allow the pathogen to escape immune detection because it is seen as *self*, or it may induce an autoimmune reaction. Because MHC molecules differ in their ability to accommodate different peptides, individuals who express certain MHC genes may lack the ability to present microbial epitopes capable of inducing protective T-cell responses. Finally, there may simply be no T cells capable of recognizing a particular MHC-antigen combination, leading to a "hole in the T-cell repertoire."

KEY POINTS ABOUT THE ASSOCIATION OF DISEASE WITH MHC HAPLOTYPE

- Certain MHC alleles are associated with a greater chance of protective immune responses to pathogens, as well as a tendency toward developing autoimmunity.
- Some diseases are distinctly more common in individuals with a particular MHC allele or MHC haplotype.

ATYPICAL AND UNIQUE ANTIGEN PRESENTATION

Superantigens Directly Bind T-Cell Receptors and MHC Without Processing

Superantigens are defined by their ability to stimulate a large fraction of T cells via interaction with the TCR Vβ domain (Fig. 5-7). Superantigens are predominantly bacterial in origin,

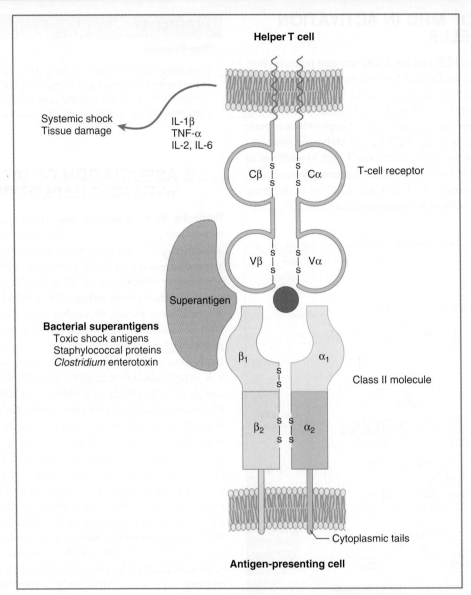

Figure 5-7. Superantigens. T cells of various antigenic specificities are activated when bacterial superantigens cross-link major histocompatability complex (MHC) class II molecules with common T-cell receptor Vβ regions. All T cells that express that particular Vβ region are subject to activation, causing massive release of cytokines and subsequent symptoms of shock and host injury. This occurs regardless of antigen specificity and the peptide that fills the MHC class II groove. *IL*, interleukin; *TNF*, tumor necrosis factor.

such as staphylococcal enterotoxin and toxin-1 responsible for toxic shock syndrome. The superantigen circumvents normal processing and directly bridges the TCR with the MHC class II molecule, causing cells to divide and differentiate into effector cells and release T-cell factors (interleukin-2, tumor necrosis factor–α, and interleukin-1β). Because the number of T cells that share Vβ domains is high (up to 10% of all T cells), large numbers of T cells (regardless of antigen specificity) may be activated by superantigens. This can lead to massive systemic disruption with symptoms similar to septic shock including occurrence of severe tissue injury and possible multiple organ failure.

Presentation of Nonclassical Lipid Antigens to T Cells

The CD1 molecules represent a novel lineage of antigen-presenting molecules. They are distinct from MHC classes I and II; the CD1 family of non-MHC locus-encoded proteins presents nonpeptide glycolipid antigens to T cells (Fig. 5-8). In humans, there are five distinct CD1 proteins (CD1a, b, c, d, and e). The CD1 molecules have a similar structure to MHC class I, with three extracellular domains and a noncovalently associated β_2M. The peptide groove is lined with nonpolymorphic (semi-invariant) TCR that allows recognition of glycolipids through the CD1d molecule.

KEY POINTS ABOUT ATYPICAL AND UNIQUE ANTIGEN PRESENTATION

- Superantigens cross-link TCRs and MHC independent of antigen and antigen processing, usually via direct interaction with the TCR Vβ region.
- The CD1 family of non-MHC locus-encoded proteins presents nonpeptide glycolipid antigens to T cells.

KEY POINTS

- Antigen is recognized in conjunction with proteins of the MHC. MHC molecules are highly allotypic and polymorphic.
- Class I region genes consist of HLA-A, HLA-B, and HLA-C loci with each gene coding a transmembrane polypeptide containing three extracellular domains (α_1, α_2, α_3). Proteins are expressed at the cell surface in a noncovalent association with β_2M. Class I molecules are found on all nucleated cells, and are critical for presentation of antigen to CD8+ CTLs.
- The class II region genes consist of the D region, which is subdivided into HLA-DP, HLA-DQ, and HLA-DR subregions. Class II molecules have two transmembrane polypeptide chains; peptides bind within the allele-specific pockets defined by chain interactions. Class II molecules, found primarily on B cells and macrophages, present antigen in the presence of cytokines to CD4+ lymphocytes.
- T cells in the thymus learn to recognize antigenic peptides in the context of self-MHC molecules by a two-step process involving both positive and negative selection.
- Different antigen degradation and processing pathways produce MHC-peptide complexes in which *endogenous* peptides associate with class I molecules and *exogenous* peptides associate with class II molecules.
- Certain MHC alleles are associated with a greater chance of protective immune responses to pathogens, as well as toward development of autoimmunity. Some diseases are distinctly more common in individuals with a particular MHC allele or MHC haplotype.
- Superantigens cross-link TCRs and MHC without antigen processing, causing a cytokine storm produced by massive and nonspecific T-cell activation.
- The CD1 molecules are similar in structure to the MHC class I protein, and present nonpeptide glycolipid antigens to T cells.

Self-assessment questions can be accessed at www.StudentConsult.com.

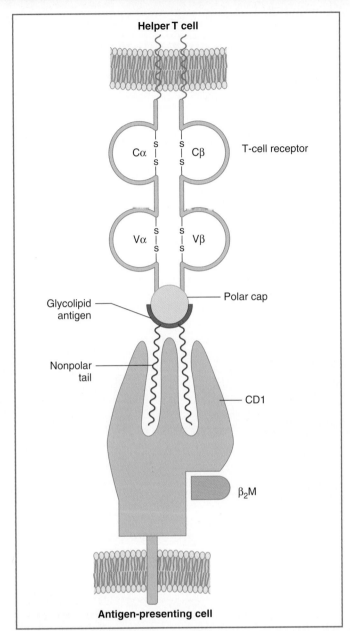

Figure 5-8. Nonclassic lipid antigen presentation by CD1 molecules. The peptide groove of the CD1 surface molecule is lined with nonpolar, hydrophobic side chains. The CD1 molecule binds antigen in a deep, narrow hydrophobic pocket, enabling presentation to the T-cell receptor. β_2M, β_2-microglobulin.

nonpolar/hydrophobic side chains in which antigen binds in a deep, narrow hydrophobic pocket; ligands interact via hydrophobic interactions rather than by hydrogen bonding. Although the role of the CD1 in pathogenesis has not been determined, it is an area of active investigation. Natural killer T (NKT) cells are a specific subset of T cells coexpressing a

Innate Immunity

CONTENTS
PHYSICAL BARRIERS AGAINST INFECTION
  Anatomic Barrier
  Physiologic Barrier
  Phagocytic and Endocytic Barriers
  Inflammatory Barriers
DEFENSIVE INNATE RESPONDING CELLS
  Neutrophils
  Mononuclear Cells and Macrophages
MICROBIAL AGENT RECOGNITION THROUGH PATTERN RECOGNITION RECEPTORS
LINKING INNATE AND ADAPTIVE IMMUNE RESPONSES
NATURAL KILLER CELLS: EARLY DEFENDERS AGAINST INFECTION
COMPLEMENT IN INFLAMMATION AND IMMUNITY
  Complement Cascades
  Biologic Functions of Complement
FEVER AND RASH

The innate immune system is composed of components that are present before the manifestation of infection. These components include physical defensive barriers as well as factors that recognize classes of molecules frequently encountered on invading pathogens. Innate defensive measures allow containment of organisms while the specific immune response is either generated or upregulated.

PHYSICAL BARRIERS AGAINST INFECTION

Four major categories of physical barriers exist to limit entry and control expansion of foreign pathogens. Defensive roles may be *anatomic* (skin, mucous membranes), *physiologic* (temperature, low pH, chemical mediators), *phagocytic* (digestion of microorganisms), or *inflammatory* (vascular fluid leakage).

Anatomic Barrier

The skin and mucous membranes provide an effective barrier against microorganisms. The skin has the thin outer epidermis and the thicker underlying dermis to impede entry as well as sebaceous glands to produce sebum. Sebum is made of lactic acid and fatty acids, which effectively reduce skin pH to between 3 and 5 to inhibit organism growth. Mucous membranes line the sites of contact between the body and the outside world. In addition to locally produced antimicrobial substances, they are covered by cilia, which function to trap organisms in mucus and propel them out of the body.

Specific organs have additional levels of innate protective responses. For example, the urinary tract has altered pH, a long urethra, and bladder flushing to prevent entry. The vagina secretes estrogens to promote an acidic environment unfavorable to pathogen growth. Milk and vaginal secretions contain lactoperoxidase, lactoferrin, and lysozymes as antimicrobial agents. Secretory immunoglobulin (Ig) A, found in the above surfaces as well as in tears, saliva, and nasal and bronchial tissues, plays a critical role in opsonization of organisms and blocking adherence of pathogens to epithelial surfaces.

Physiologic Barrier

The physiologic barrier includes factors such as temperature, low pH, and chemical mediators. Many organisms cannot survive or multiply in the presence of elevated body temperature. Soluble proteins such as lysozymes, interferons, and complement components play a major role in innate immunity. Lysozymes can interact with bacterial cell walls; interferons α and β are natural inhibitors of viral growth; complement components use both specific and nonspecific immune factors to convert inactive forms to active moieties that damage membranes of pathogens. Low pH in the stomach and gastric environment discourages bacterial growth.

Phagocytic and Endocytic Barriers

Blood monocytes, tissue macrophages, and neutrophils phagocytose and kill microorganisms via multiple complex digestion mechanisms. Bacteria become attached to cell membranes and are ingested into phagocytic vesicles. Phagosomes fuse with lysosomes where lysosomal enzymes digest captured organisms.

HISTOLOGY

The Skin

The skin is composed of a thin outer epidermis and thicker underlying dermis and assists in impeding entry of foreign organisms. The interface between the two tissues is demarcated by epidermal ridges and dermal papillae, with interpapillary pegs dividing ridges. Production of sebum (lactic acid and fatty acids) through short ducts emanating from sebaceous glands effectively reduces skin pH, which inhibits the growth of microorganisms.

Inflammatory Barriers

Invading organisms cause localized tissue damage leading to complex inflammatory responses (Fig. 6-1). Initial damage and entry of organisms result in release of chemotactic factors (complement components, chemokines, fibrinopeptides) to signal changes in nearby vasculature, allowing extravasation of polymorphonuclear cells to the injury site. Celsus (first century AD) described the four cardinal signs of inflammation as *rubor* (redness), *tumor* (swelling), *calor* (heat), and *dolor* (pain). Later, Galen (second century AD) added a fifth sign: *functio laesa* (loss of function).

KEY POINTS ABOUT PHYSICAL BARRIERS

- Innate defense barriers include (1) anatomic barriers, (2) physiologic barriers, (3) phagocytic barriers, and (4) inflammatory barriers.
- Damage to tissue caused by invading pathogens can lead to *rubor, tumor, calor, dolor,* and *functio laesa.*
- Tissue damage leads to an influx of inflammatory cells through chemotaxis, activation, margination, and diapedesis.

PATHOLOGY

Acute Respiratory Distress Syndrome

Acute respiratory distress syndrome is characterized by lung inflammation caused by excessive nitric oxide and inflammatory cascades, leading ultimately to vasodilatation and fluid exudation. Congestion of the vessels and proteinaceous acellular edema occur in the alveoli with accumulation of vividly eosinophilic, acellular hyaline membranes against alveolar septae.

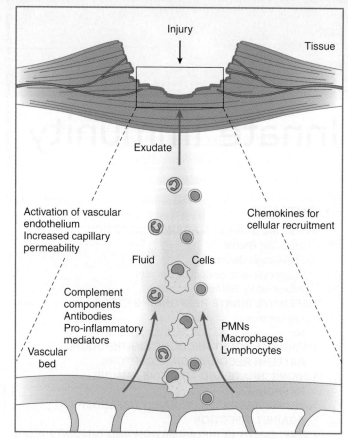

Figure 6-1. Inflammatory responses lead to vasodilation, causing erythema (redness) and increased temperature, and increased capillary permeability, which allows exudates (fluid) to accumulate, leading to tissue swelling (edema) and influx of cells to the site of tissue damage. Once cells enter the area of injury, they release chemotactic factors to recruit additional cells, leading to local activation at the damaged site.

●●● DEFENSIVE INNATE RESPONDING CELLS

The cell types involved in innate immune responses include the polymorphonuclear cells (primarily neutrophils), monocytes and macrophages, eosinophils, and natural killer (NK) cells. Some of these cells are capable of killing target cells via nonspecific (nonantigen/major histocompatability complex [MHC] dependent) means through release of lytic enzymes, perforin, or tumor necrosis factor (TNF). Others are involved in phagocytic mechanisms that kill via intracellular processes.

Neutrophils

Neutrophils are typically the first infiltrating cell type to the site of inflammation (Fig. 6-2). Activated endothelial cells increase expression of E-selectin and P-selectin, which are recognized by neutrophil surface mucins (PSGL-1 or sialyl Lewis^X on glycoproteins or glycolipids), and induce neutrophil rolling along the endothelium. Chemoattractants such as interleukin (IL)-8 or CXCL8 can trigger firm adhesion and diapedesis utilizing molecules such as leukocyte function-associated antigens (e.g. LFA-1 and LFA-3). Subsequent chemotaxis is further induced by multiple complement components (e.g., C5a), fibrinopeptides, and leukotrienes. Activated neutrophils express high-affinity Fc receptors and complement receptors to allow increased phagocytosis of invading organisms. Activation of neutrophils leads to the respiratory burst, producing reactive oxygen and nitrogen intermediates as well as release of primary and secondary granules containing proteases, phospholipases, elastases, and collagenases. Neutrophils also release cationic defensins, lactoferrin, and myeloperoxidases that assist in both direct antipathogenic functions and initiation of secondary immune responses. Pus, a yellowish-white, opaque, creamy matter produced by the process of suppuration, consists of innumerable neutrophils (some dead and dying) and tissue debris.

Mononuclear Cells and Macrophages

Macrophages involved in innate immunity include **alveolar macrophages** in the lung, **histocytes** in connective tissue, **Kupffer cells** in the liver, **mesangial cells** in the kidney, **microglial cells** in the brain, and **osteoclasts** in bone. Each subset phenotype is unique in its expression levels of specfic surface molecules, such as chemokine or Fc (immunoglobulin)

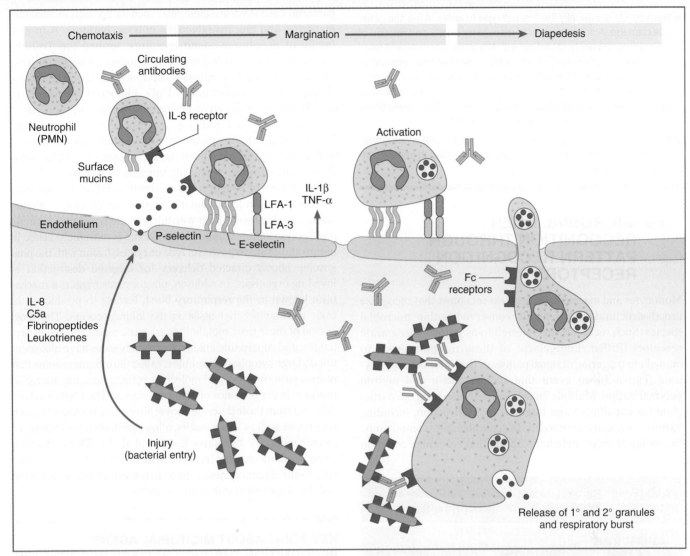

Figure 6-2. Events associated with neutrophil transendothelial migration. Bacteria entering through a breach in the mechanical barrier (skin) trigger release of chemotactic factors that upregulate selectins on endothelial beds. Circulating polymorphonuclear (*PMNs*) cells interact via weak binding surface mucins; interactions are enhanced in the presence of stimulating factors (interleukin [IL]-1, IL-8, tumor necrosis factor (*TNF*)-α), leading to further cellular activation, margination, and diapedesis. Pictured are neutrophils, which, when entering tissue, undergo respiratory burst and release of primary and secondary granules upon cross-linking of Fc receptors to antibodies recognizing bacterial epitopes. *LFA*, leukocyte function-associated antigen.

receptors. Chemokine mediators such as macrophage inflammatory protein-1α (MIP-1α) and MIP-1β attract monocytes to the site of pathogenic infection (Table 6-1). Like neutrophils, monocytes express surface ligands (e.g., integrin a4b1, also called VLA-4) that recognize ligands (e.g., vascular cell adhesion molecule-1 [VCAM-1]) on endothelial cells. This interaction can mediate cell rolling, firm adhesion, and diapedesis. Activated tissue macrophages secrete proinflammatory mediators in response to bacteria and bacterial products, including IL-1, IL-6, IL-8, IL-12, and TNF-α. TNF-α is an inducer of a local inflammatory response that helps contain infections. IL-8 also is involved in the local inflammatory response, helping attract neutrophils to the site of infection. IL-1, IL-6, and TNF-α have a critical role in inducing the acute-phase response in the liver and inducing fever, which limits bacterial growth and upregulates host defense in several ways. IL-12 may also activate natural killer cells, another early innate cell phenotype.

KEY POINTS ABOUT DEFENSIVE INNATE RESPONDING CELLS

- Immune responses of the innate immune system provide natural immunity against microorganisms via phagocytosis and intracellular killing, recruitment of other inflammatory cells, and presentation of antigens. Leukocytes that provide innate immunity are derived from **myeloid** lineage. These cells include highly phagocytic, motile neutrophils; monocytes and tissue macrophages; eosinophils; and NK cells. These cells provide the first line of defense against most pathogens.

Continued

- Neutrophils are usually the first cell type to arrive at the site of tissue damage. Activation leads to respiratory bursts and release of granules to control bacterial growth. Mononuclear cells and macrophages engulf organisms via multiple mechanisms, leading to control and destruction within intracellular phagosomes. NK cells are large granular lymphocytes that kill targets via antibody-dependent cell-mediated cytotoxicity or through lysis using either Fas-mediated or perforin-induced mechanisms.

- Chemokines and complement components are critical for activation of innate immune functions. Defects may lead to severe clinical complications.

MICROBIAL AGENT RECOGNITION THROUGH PATTERN RECOGNITION RECEPTORS

Monocytes and macrophages express receptors that recognize broad structural motifs highly conserved within microbial species. Such receptors are referred to as *pattern recognition receptors* (PRRs). Engagement of these receptors leads to immediate triggering of signal pathways that promote phagocytosis (Fig. 6-3), an event that requires actin and myosin polymerization. Multiple factors assist in preparing the particulate for engulfment and targeting for destruction, including various opsonins composed of complement components. Examples of these include receptors that recognize common bacterial carbohydrate elements, such as lipopolysaccharide, mannose, and glucans (glycogen polysaccharides). A unique family of membrane-bound receptors, termed the *Toll-like receptors*, plays a critical role in recognition of bacterial components (cell wall membranes, lipopolysaccharide, flagellin, nucleic acids, lipopeptides, CpG oligodeoxynucleotides); specificity of engagement occurs via interaction with either homodimer or heterodimer receptor configurations that initiate local and/or systemic proinflammatory responses. Toll-like receptors have been shown to regulate both the magnitude and duration of ensuing immune responses.

The process of phagocytosis compartmentalizes the invading pathogen into an intracellular vacuole referred to as a **phagosome.** The phagosome fuses with intracellular lysosomes, forming a phagolysosome vacuole. The lysosome is extremely acidic in nature and contains powerful lytic enzymes; fusion with the phagosome allows directed delivery for targeted destruction of invading organisms. In addition, phagocytosis triggers a mechanism known as the **respiratory burst,** leading to production of toxic metabolites that assist in the killing process. The most critical of these toxic metabolites are nitric oxide, hydrogen peroxides, and superoxide anions. Recent advances have also identified a large cytoplasmic complex, called the inflammasome, that utilizes pattern-associated molecular patterns to sense microbial products in the presence of "danger" signals. The PRRs used are different from those describe above; interaction with leucine-rich repeats, as well as other motifs, allows proteolytic activation of proinflammatory cytokines IL-1β and IL-18. This inflammasome-based mechanism has been linked to development of multiple forms of pathogenesis related to protection against infection and development of autoimmune disease states.

TABLE 6-1. Partial List of Chemokines Produced by Monocytes and Macrophages

CHEMOKINE CLASS*	CHEMOKINE	CELLS AFFECTED
CC	MIP-1α	Monocytes, NK cells, T cells, dendritic cells
	MIP-1β	Monocytes, NK cells, T cells, dendritic cells
	MCP-1	Monocytes, NK cells, T cells, dendritic cells
	Eotaxin	T cells, eosinophils
CXC	IL-8	Neutrophils (naive T cells)
	GROα, β, γ	Neutrophils (naive T cells)
	IP-10	Monocytes, NK cells, T cells

*In general, chemokines fall mainly into two distinct groups. The *CC chemokines* have two adjacent cysteine residues (hence the name *CC*). The *CXC chemokines* have an amino acid between two cysteine residues. Monocytes and macrophages produce small polypeptides that are chemotactic for leukocytes, acting through receptors that are members of the G-protein–coupled signal-transducing family. All chemokines are related in amino acid sequence and their receptors are integral membrane proteins that are characterized by containing seven membrane-spanning helices. Each chemokine reacts with one or more receptors and can affect multiple cell types.
MIP, macrophage inflammatory protein; *MCP,* monocyte chemotactic protein; *IL,* interleukin; *GRO,* growth-related protein; *IP-10,* interferon-inducible protein 10; *NK,* natural killer.

KEY POINT ABOUT MICROBIAL AGENT RECOGNITION THROUGH PATTERN RECOGNITION RECEPTORS

- PRRs present on innate immune system cells assist in the recognition of bacteria and virions. Recognition by PRRs leads to activation of multiple facets of cellular response.

LINKING INNATE AND ADAPTIVE IMMUNE RESPONSES

Phagocytosed or pinocytosed antigens processed by antigen-presenting cells may then be presented to the adaptive immune system cells. Dendritic cells (DCs), macrophages and monocytes, and B cells are extremely good at presenting antigens to T lymphocytes. Indeed, the CD11c+ DCs are potent antigen-presenting cells that dictate progression of subsequent immune function, especially in the context of pathogenic invasion. The context in which the T cell recognizes antigen, the abundance of antigen, and the duration of antigen exposure are important parameters that affect the nature and maturation of the T-cell response. Most DCs are found in an immature state. Upon maturation and priming with antigen, they convert to cells that contain nonlysosomal vesicles and express large amounts of

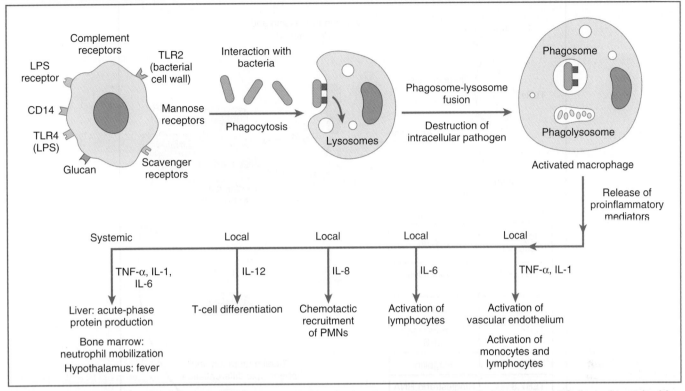

Figure 6-3. Phagocytosis by macrophages leads to activation and pathogen destruction. *IL*, interleukin; *LPS*, lipopolysaccharide; *PMN*, polymorphonuclear cell; *TLR*, Toll-like receptor; *TNF*, tumor necrosis factor.

peptide-loaded MHC molecules on their surface. The DCs then migrate to secondary lymphoid organs, such as the spleen and lymph nodes, to present antigenic fragments to naive CD4+ T cells and CD8+ cytotoxic T cells. The DCs are the most effective activators of naive-T cells. These dendritic populations have high expression of both MHC and costimulatory molecules, which are required for maximal T-cell stimulation (Fig. 6-4).

> **HISTOLOGY**
>
> **Cytoskeletal Remodeling During Phagocytosis**
>
> The cellular cytoskeleton exists as a three-dimensional network of fibrils to form a supporting scaffold to anchor structures to the plasma membrane or membrane-bound proteins, or to serve as "tracks" for organelle intracellular movement. Phagocytosis of microorganisms involves remodeling of the macrophage cytoskeleton, with regulated changes to lamellipodia. Receptor-mediated endocytosis involves engulfment of particulates with fission into membrane-bound vesicles via clathrin-coated pits; vesicle membranes are recycled to the plasma membrane, while the internalized materials are delivered to lysosomal compartments.

●●● NATURAL KILLER CELLS: EARLY DEFENDERS AGAINST INFECTION

NK cells are large granular lymphocytes that nonspecifically kill virus-infected cells and tumor cells. NK cells are effective regulators of immune function. Killing by NK cells is enhanced by cytokines such as IFN-α, IFN-β, and IL-12, and further activation can occur in the presence of activated T cells. NK cells may be activated by microorganisms to produce cytokines (IL-2, IFN-γ, IFN-α, and TNF-α). These circulating large granular lymphocytes do not express CD3, TCR, or immunoglobulin but display surface receptors (CD16) for the Fc fragment of IgG antibodies. The NK cells are able to kill "self" in the absence of antigen-specific receptors. They can also kill by means of **antibody-dependent cell-mediated cytotoxicity** mechanisms via their Fc receptors (Fig. 6-5). NK cells mediate lysis target cells by release of lytic granules and perforin-induced pore formation.

In addition, NK cells can kill target cells using Fas/Fas ligand (FasL)–mediated apoptotic pathways. NK cells bear FasL on their surface and can kill Fas-bearing target cells. NK cells do not express specific antigen receptors. Rather, NK cells use two categories of receptors to deliver either activation or inhibition signals. These opposing signals mediate activity. Activation receptors (carbohydrate-binding proteins) are always present on the surface of NK cells. Engagement of inhibitory signals (KIR and/or CD94) on the surface of the NK cell with class I MHC on the target cell counteracts activation. During viral infection, the level of class I MHC decreases on the infected target, thus allowing NK cells to be fully activated to destroy the target cell.

●●● COMPLEMENT IN INFLAMMATION AND IMMUNITY

Complement is a system of more than 30 serum and cell-surface proteins involved in inflammation and immunity. In conjunction with specific antibodies, complement components act as

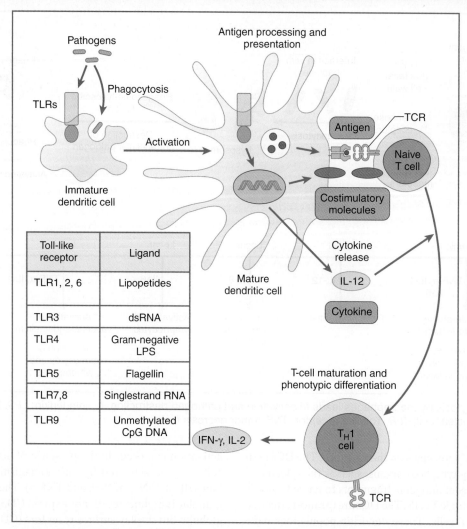

Figure 6-4. Link between innate and adaptive (acquired) immunity. Pathogen recognition through pattern recognition receptors is an important bridge between innate and adaptive immune function. Recognition leads to activation and maturation of the presenting cell. Here, dendritic cells are depicted as primary presenting cells, which assist in dictating subsequent responses. Processed antigen is presented to naive T cells, accompanied by secretion of cytokines to assist development and maturation of T-cell phenotypic response (e.g., T helper cell-1 [T_H1] maturation via presence of interleukin [IL]-12). Inset box shows important Toll-like receptors (*TLRs*) and specific ligands involved in pathogen recognition. At least 15 different *TLRs* have been identified, with ligand motifs identified for most of them. *TCR*, T-cell receptor; *LPS*, lipopolysaccharide; *IFN*, interferon.

the primary humoral defense system against bacterial and viral infections. Most of the complement proteins in the serum are produced by liver hepatocytes. Some of the complement components are acute-phase proteins and can increase in concentration by twofold to threefold. Many of the complement proteins have shared sequences, indicating that they evolved by gene duplication and recombination.

Complement Cascades

Complement activation involves the sequential activation of complement proteins, either by protein-protein interactions or by proteolytic cleavage. At each step, the number of protein molecules activated increases, amplifying the reaction. Many complement proteins are present as **zymogens** (inactive precursors), which are activated either by conformational changes or by proteolytic cleavage by other complement proteins. Activation of these zymogens results in specific serine protease activity capable of cleaving other complement proteins, producing the complement cascade. Three pathways are described for complement activation. The **classic pathway** is initiated by the presence of antigen-antibody complexes. The **alternative pathway** is initiated by recognition of foreign cell surfaces. The **lectin pathway** (or **mannose-binding pathway**) recognizes mannose on pathogenic organisms (Fig. 6-6).

The classical complement pathway is initiated by immunocomplexes of antigen and antibody. Complement factor C1 (composed of C1q, C1r, and C1s) is activated through requisite cross-linking of two antibodies through Fc region binding. Alternatively, one pentameric IgM molecule may serve as the catalyst. Activated C1 complex can catalyze the cleavage of C4, followed by C2, yielding C4b2b, which is known as the *C3 convertase*. The C3 convertase amplifies breakdown of C3 into C3a and C3b; C3a is a powerful chemotactic factor

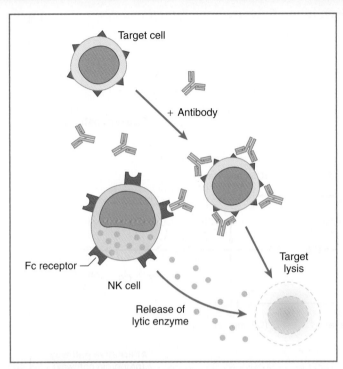

Figure 6-5. Antibody-dependent cell-mediated cytotoxicity (ADCC). ADCC is a phenomenon by which target cells coated with antibody are destroyed by specialized killer cells. Among the cells that mediate ADCC are natural killer (NK) cells, neutrophils, and eosinophils. The killing cells express receptors for the Fc portion of antibodies that coat targets Recognition of antibody-coated targets leads to release of lytic enzymes at the site of Fc-mediated contact. Target cell killing may also involve perforin-mediated membrane damage. Eosinophils can function in a similar manner to kill large parasites.

that can also increase vascular permeability. C3b joins with the C4b2b complex to form the C5 convertase. Breakdown of C5 leads to production of C5a, another powerful chemotactic factor and macrophage activator. C5b forms the basis for creation of the membrane attack complex (MAC), which, in essence, assembles pore structures within cell membranes, leading to hypotonic lysis of target cells. Whereas the small C5a fragment is released into the blood as a potent complement anaphylatoxin, the large C5b molecule binds proteins C6 and C7. The complex C5b67 has hydrophobic regions that permit it to insert into the lipid bilayer of nearby cell membranes. Subsequent binding of C8 permits some leakage of cell contents, causing slow lysis. This process is accelerated by binding of multiple C9 molecules, which assemble to form a protein channel through the membrane. C9 is analogous to perforins produced by cytolytic T cells and NK cells. C5b6789 thereby forms the basis of the MAC; the MAC is especially critical for elimination of bacteria resistant to intracellular killing by phagocytes.

The classic pathway of complement can also be activated by a serum mannose-binding lectin (MBL) complex. This complex is structurally similar to the C1 complex. However, instead of binding to immunocomplexes, it binds directly to polysaccharides on gram-negative bacteria. The MBL is C1q-like in structure, and the mannose-associated serine proteases (MASPs) are similar to C1r and C1s. Upon binding bacterial surfaces, MBL-MASP can cleave C4 and C2 and thereby activate the remainder of the classical pathway.

The complement component C3 is the starting point for the alternative pathway. Cleavage of C3 exposes a reactive thioester bond that enables C3b to bind to bacterial cell surfaces. Factor B is bound, and the complex is cleaved by factor D, resulting in **C3 convertase**. C3 convertase allows amplification of enzymatic cleavage of C3 into more C3a and C3b. Factor H displaces B from C3bBb and further assists in the conversion of C3b to iC3b by acting as a cofactor for factor I. C3a can also bind to mast cells to cause release of histamines. iC3b acts to activate neutrophils and macrophages via CR3 receptors (in the absence of antibody).

Biologic Functions of Complement

Activation of the complement system results in the production of several different polypeptide cleavage fragments that are involved in five primary biologic functions of inflammation and immunity (Fig. 6-7).

Direct Cytolysis of Foreign Organisms

Antibodies recognizing pathogenic determinants form the basis of a physical structure with which complement components interact. Specifically, complement component C1 interacts with the Fc portion of IgM and IgG (except IgG4), binding to the surface of bacteria. The binding of C1 initiates a cascade of events whereby a MAC is built upon the cellular surface. Synthesis of the MAC structure culminates in assembly of a pore channel in the lipid bilayer, causing osmotic lysis of the cell. MAC formation requires prior activation by either the classical or the alternative pathway and utilizes the proteins C5b, C6, C7, C8, and C9.

Opsonization of Foreign Organisms

Complement components (e.g., C3b or inactivated C3b; iC3b) bind to pathogens. Interaction with receptors (CR1, CR2, CR3, and CR4) on the surface of macrophages, monocytes, and neutrophils leads to enhanced phagocytosis and targeted destruction of organisms.

Activation and Directed Migration of Leukocytes

Proteolytic degradation of C3 and C5 leads to production of leukocyte chemotactic factors referred to as anaphylatoxins. For example, C3a is chemotactic for eosinophils. C5a is a much more potent chemokine, attracting neutrophils, monocytes and macrophages, and eosinophils. Interaction of C3a, C4a, or C5a with mast cells and basophils leads to release of histamine, serotonin, and other vasoactive amines, resulting in increased vascular permeability, causing inflammation and smooth muscle contraction.

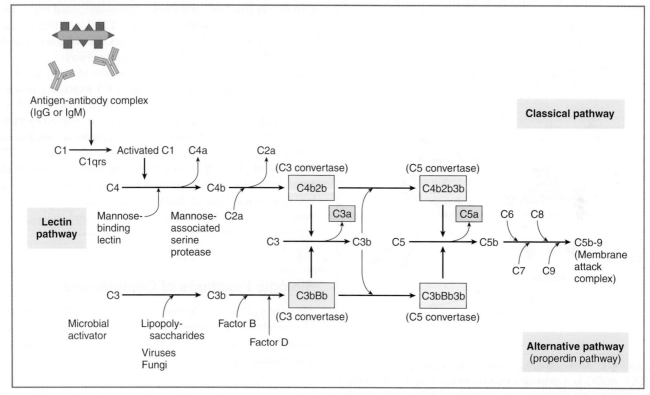

Figure 6-6. Activation of complement through the *classic pathway* (antigen-antibody complexes), the *alternative pathway* (recognition of foreign cell surfaces), or the *lectin pathway* (or *mannose-binding pathway*) promotes activation of C3 and C5, leading to construction of the membrane attack complex. *Ig*, immunoglobulin.

Solubilization and Clearance of Immunocomplexes

One of the major roles complement plays is the solubilization and clearance of immunocomplexes from the circulation. First, C3b and C4b can covalently bind to the Fc region of insoluble immunocomplexes, disrupting the lattice and making them soluble. C3b and C4b bound to the immunocomplex are recognized by the CR1 receptor on erythrocytes, facilitating their transport to the liver and spleen. In the liver and spleen, the immunocomplexes are removed and phagocytosed by macrophage-like cells. The red blood cells are returned to the circulation.

Enhancement of Adaptive Immune Function

Complement factors are traditionally known as an effector arm of humoral immunity. Indeed, coating of antigens with C3d (a breakdown product of C3) facilitates their delivery to germinal centers rich in B cells and follicular DCs. Recent discoveries also highlight the importance of complement to regulate T-cell response, including requirements for T-cell lineage development. It is now appreciated that many complement factors bridge both innate and adaptive functions.

KEY POINTS ABOUT COMPLEMENT IN INFLAMMATION AND IMMUNITY

- The activities of complement include (1) cytolysis of foreign organisms, (2) opsonization and phagocytosis of foreign organisms, (3) activation of inflammation and directed migration of leukocytes, (4) solubilization and clearance of immunocomplexes, and (5) enhancement of humoral and adaptive immune response.

- Complement activation results in the release of anaphylatoxins (C3a, C4a, and C5a). These are important mediators of inflammation, causing recruitment and activation of neutrophils, macrophages, and other cell types.

- Activation also produces cleavage products (C3b, C3bi, and C4b), which serve as opsonins, enhancing phagocytosis.

●●● FEVER AND RASH

The initial outcome of a strong proinflammatory response can often lead to secondary outcomes of fever and rash. Fever, with coinciding myalgias, arthralgias, and headaches, is usually self-limiting when organisms are cleared. It is important to understand that fever is not directly caused by pathogenic factors; rather, these factors trigger physiologic mechanisms controlled by the hypothalamus. The pyrogenic cytokines, TNF-α (cachectin), TNF-β (lymphotoxin), IL-1, and IL-6 are potent inducers of fever response. Fever is beneficial to the host in that most pathogens replicate poorly in elevated systemic temperatures.

Rashes occur for a variety of reasons; however, they too can be attributed to innate response. Local detection of pathogens and complement-mediated lysis of organisms lead to release of proinflammatory mediators TNF-α and IL-1β that initiate a local hypersensitive effect. Other mechanisms involve direct lysis of infected cells, whether by complement and antibody and resultant MAC formation, by NK activity, or by subsequent adaptive T-cell–mediated specific targeted cell destruction.

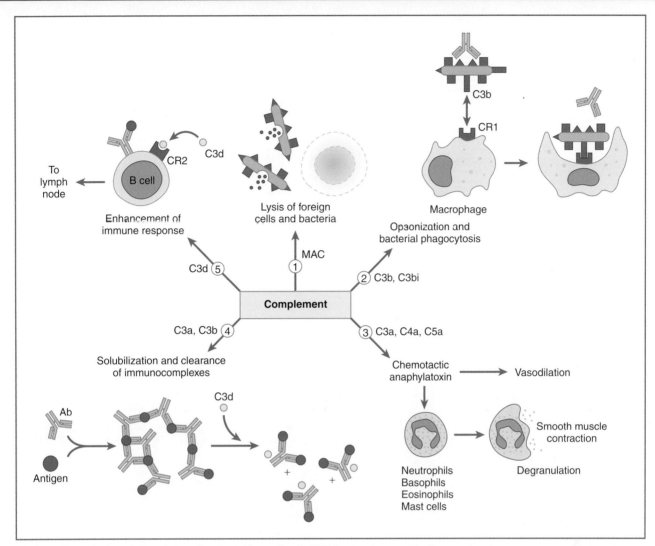

Figure 6-7. Biologic functions of complement. *Ab*, antibody; *MAC*, membrane attack complex.

In many cases, fever and rash are also accompanied by lymphadenopathy resulting from excessive draining of immune factors, mediators, and activated cells to surrounding lymph nodes.

KEY POINTS

- Tissue damage and resultant inflammation occur due to mechanisms controlling chemotaxis, activation, margination, and subsequent diapedesis of polymorphonuclear cells. The eventual outcome can be described as the four cardinal signs: *rubor* (redness), *tumor* (swelling), *calor* (heat), and *dolor* (pain), and sometimes *functio laesa* (loss of function).

- Serum complement components and antibody reactivity are critical for activation of innate immune cascades.

- Myeloid leukocytes provide innate immunity's first line of defense against pathogens. Neutrophils are rapid responders, employing respiratory bursts and release of cidal granules to slow organism growth. Monocytes and tissue macrophages, eosinophils, and NK cells also contribute to the proinflammatory response.

- Pattern-associated molecular patterns present on invading organisms dictate responses, allowing directed progression to development of protective immunity.

- Complement function may be divided into the classical, lectin, and alternative (properdin) pathways, culminating in actions leading to (1) cytolysis of foreign organisms, (2) opsonization and phagocytosis of foreign organisms, (3) activation of inflammation and directed migration of leukocytes, (4) solubilization and clearance of immunocomplexes, and (5) enhancement of adaptive immune function.

Self-assessment questions can be accessed at www.StudentConsult.com.

Adaptive Immune Response and Hypersensitivity

CONTENTS
T CELLS: PRIMARY EFFECTORS FOR CELL-MEDIATED IMMUNITY
DEVELOPMENT OF T HELPER 1 AND 2 CELL PHENOTYPES
ROLE OF T CELLS IN B-CELL ACTIVATION
DEVELOPMENT OF CYTOTOXIC T-CELL EFFECTORS
HYPERSENSITIVITY: INAPPROPRIATE RESPONSE RESULTING IN DISEASE
   Type I Hypersensitivity: Immunoglobulin E–Mediated Immediate Hypersensitivity
   Type II Hypersensitivity: Antibody-Mediated Hypersensitivity
   Type III Hypersensitivity: Immunocomplex-Mediated Hypersensitivity
   Type IV Hypersensitivity: Delayed-Type (Cell-Mediated) Hypersensitivity

T CELLS: PRIMARY EFFECTORS FOR CELL-MEDIATED IMMUNITY

The T cell is the primary effector for control and regulation of adaptive immune function. T helper 1 (T_H1) cells aid in the regulation of cellular immunity. T helper 2 (T_H2) cells aid B cells in isotype switching and in production of antibodies. Cytotoxic T cells (CTLs) mediate antigen-specific, major histocompatibility complex (MHC)-restricted cytotoxicity and are important for killing intracytoplasmic pathogens that are not accessible to secreted antibody or to phagocytes.

KEY POINTS ABOUT T CELLS

- Adaptive immune responses are mediated by effector lymphocytes.
- T_H cells provide assistance in regulation of cellular immunity and help to B cells for isotype switching and production of antibodies.
- CTLs mediate antigen-specific, MHC class I–restricted cytotoxicity and killing of intracellular pathogens.

DEVELOPMENT OF T HELPER CELL PHENOTYPES

The development of T helper cell and T helper cell cellular phenotypes (defined primarily by patterns of cytokine expression) is under the control of local cytokines and factors (Fig. 7-1). Development of precursor T cells into the T_H1 phenotype gives rise to cells that promote cytotoxic and delayed-type hypersensitivity reactions as well as activate macrophages for increased protection against intracellular pathogens. The T_H1 cells are primarily characterized by secretion of interferon-γ and interleukin (IL)-2. T_H2 cells primarily secrete IL-4, IL-5, IL-6, and IL-10 and activate B cells for antibody production. Each phenotype produces cytokines that inhibit development of the alternate set; interferon-γ inhibits T_H2 development, whereas IL-4 and IL-10 inhibit T_H1 development. In addition, precursor cells exhibit specific production of transcription factors (STAT4 or STAT6 and GATA3) that influence lineage development. Maturation of other T cell phenotypes (as discussed in chapter 4) are similarly under control of local signals and cytokines present during development.

KEY POINTS ABOUT THE DEVELOPMENT OF T HELPER 1 AND 2 PHENOTYPES

- The development of T_H response is dictated by local production of cytokines, giving rise to T_H1 or T_H2 cells from naive precursor cells.
- Each subset produces a unique set of cytokine mediators, some of which inhibit development of the other phenotype.

ROLE OF T CELLS IN B-CELL ACTIVATION

B cells are capable of presenting antigen, thus allowing direct elicitation of cytokines from T cells for stimulation, resulting in activation and plasma cell development (Fig. 7-2). Clonal expansion of B cells under the influence of T-cell cytokines leads to plasma cell development, isotype class switching, and production of memory responses. In general terms, the presence of local

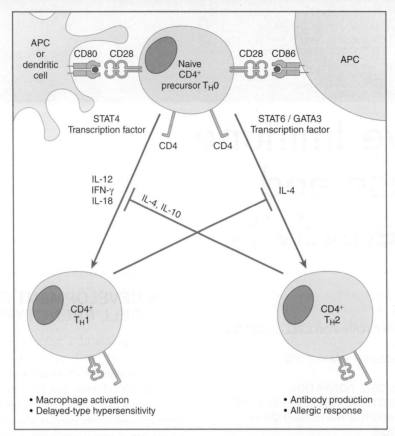

Figure 7-1. Development of T helper (T_H) cell phenotypes. T_H1 and T_H2 cells derive from precursor cells (T_H0) under the influence of local cytokines. Each secretes a phenotypic subset of cytokines that drive effector responses. In addition, secreted cytokines modulate response to the other subset and inhibit alternative functional development. Not shown are the T_H17 and T regulatory subsets, which mature under the influence of cytokine interleukin (*IL*)-17 and tumor growth factor-β. *APC*, antigen-presenting cell.

cytokines will dictate isotype switching. For example, IL-4 secreted from T_H2 cells acts as a B-cell growth factor, and IL-6 assists in delivering signals for maturation of the antibody response. Under certain circumstances, B cells can also respond and proliferate through T-independent mechanisms, usually involving antigens with long repeating epitopes that allow cross-linking of immunoglobulin receptors on the surface of the B cell. A common example of this occurs with bacterial capsid antigens containing repeating carbohydrate (polysaccharide) epitopes. In the case of T-independent activation, there is no accompanying maturation of response, and antibody production is, primarily limited to immunoglobulin M production.

DEVELOPMENT OF CYTOTOXIC T-CELL EFFECTORS

The CD8+ CTL recognizes antigen presented via class I MHC molecules. These require strong costimulatory signals to become effector cells (Fig. 7-3). CD8+ T cells can be activated by peptides presented by dendritic cells, which deliver a strong costimulatory signal through surface expression of molecules such as CD80 and CD86, which bind T-cell CD28, and intercellular adhesion molecule-1 (CD54), which binds the T-cell integrin CD11a/CD18. Other antigen-presenting cells do not produce strong enough costimulatory signals; IL-2 from nearby CD4+ T_H1 cells recognizing peptide presented by class II MHC on the same antigen-presenting is required to fully activate these CD8+ CTLs to become cytotoxic effectors.

HYPERSENSITIVITY: INAPPROPRIATE RESPONSE RESULTING IN DISEASE

Hypersensitivity is a definable immune response that leads to deleterious host reactions rather than protection against disease. The hypersensitivity reactions fall into four classes based on their mechanisms and the ability to passively transfer response through antibodies or through T lymphocytes. These responses include inappropriate antigenic response, excessive magnitude of response, prolonged duration of response, and innocent-bystander effect reactions leading to tissue damage.

Type I Hypersensitivity: Immunoglobulin E–Mediated Immediate Hypersensitivity

Type I hypersensitivity (also called **immediate hypersensitivity**) is due to aberrant production and activity of IgE against normally nonpathogenic antigens (commonly called **allergens**) (Fig. 7-4). Common antigenic allergens include animal dander, chemical additives, foods, insect stings, pollens, and even drugs. Antigen presentation to T_H2 cells leads to IgE isotype

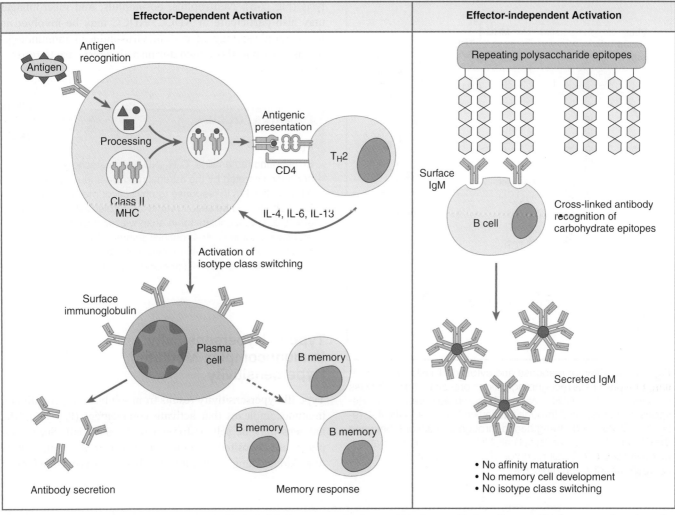

Figure 7-2. B-cell response to T-dependent and T-independent antigens. T helper 2 (T_H2) cells specifically recognize class II–presented antigenic determinants and drive B-cell activation, leading to isotype class switching and antibody secretion (*left*). Alternatively, T-independent responses to repeating carbohydrate epitopes stimulate antibody production but do not lead to maturation of the antibody response. *MHC*, major histocompatability complex; *IL*, interleukin; *Ig*, immunoglobulin.

production. The IgE binds to mast cells or basophils via high-affinity IgE receptors. Subsequent antigen exposure results in cross-linking of cell-bound IgE with activation (degranulation) of mast cells or basophils to release preformed mediators (e.g., vasoactive amines, histamine, leukotrienes, prostaglandin D_2) and to synthesize new mediators (i.e., chemotaxins, cytokines). These mediators are responsible for the signs and symptoms of allergic diseases. In severe cases, anaphylactic shock may occur, characterized by a sudden and sharp drop in blood pressure, urticaria, and breathing difficulties caused by exposure to a foreign substance (such as bee venom or drug reactivity). Emergency treatment includes epinephrine injections, used as a heart stimulant, vasoconstrictor, and bronchial relaxant.

Type II Hypersensitivity: Antibody-Mediated Hypersensitivity

Type II hypersensitivity reactions are due to antibodies directed against cell membrane–associated antigen. The end result of the antibody response is cytolysis. The mechanism may involve complement (**cytotoxic antibody**) or effector lymphocytes that bind to target cell–associated antibody and effect cytolysis via a complement-independent pathway. In complement-mediated type II hypersensitivity, antibody recognition of cell-surface epitopes leads to assembly of the complement membrane attack complex (C5-C9) and subsequent lysis of the cell (Fig. 7-5). This reaction is the underlying mechanism in multiple disease states, including that seen in autoimmune hemolytic anemia, Goodpasture syndrome (directed against basement membrane molecules), and Rh incompatibility leading to erythroblastosis fetalis. In rare cases, pharmaceutical agents, such as penicillin or chlorpromazine, can bind cells, forming a novel antigenic surface complex that provokes antibody production and type II cytotoxic reactions.

A second mechanism for type II reactions is characterized by antibody-dependent cell-mediated cytotoxicity (ADCC) induced by natural killer (NK) cells recognizing IgG attached to target cells bearing antigen. The constant portion of the antibody (Fc region) is bound by Fc receptors on the NK cell, leading to perforin release and NK cell–mediated

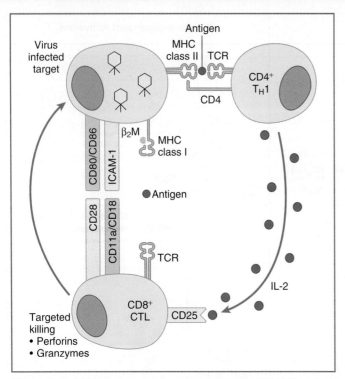

Figure 7-3. Effector response and cytotoxic T cell (*CTL*) activation. T helper 1 cells recognizing antigen presented through class II molecules drive CD8+ cells to become activated CTLs. Responses are mediated through interleukin-2 (*IL-2*) and the IL-2 receptor (CD25), and strengthened through interactions between the CTL and the infected target cell. *MHC*, major histocompatability complex; *TCR*, T-cell receptor; β_2M, β_2-microglobulin; *ICAM*, intercellular adhesion molecule.

lysis (Fig. 7-5). Neutrophils, eosinophils, and macrophages may also participate in ADCC. ADCC may be involved in the pathophysiology of certain virus-induced immunologic diseases, such as those seen during active response to retroviral infection.

PHARMACOLOGY

Epinephrine

Epinephrine (adrenaline, $C_9H_{13}NO_3$) is a catecholamine hormone secreted by the adrenal medulla and released into the bloodstream in response to physical or mental stress. Epinephrine may be used to stimulate cardiac action in cardiac arrest, as a vasoconstrictor in anaphylactic shock, and as a bronchodilator in acute bronchial asthma. Chemically, epinephrine is a sympathomimetic monoamine derived from the amino acids phenylalanine and tyrosine.

Type III Hypersensitivity: Immunocomplex-Mediated Hypersensitivity

Type III hypersensitivity results from soluble antigen-antibody immunocomplexes that activate complement (Fig. 7-6). The antigens may be self or foreign (i.e., microbial). Such complexes are deposited on membrane surfaces of various organs (e.g., kidney, lung, synovium). The by-products of complement

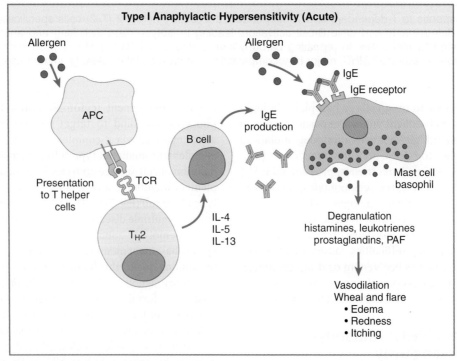

Figure 7-4. Type I acute anaphylactic hypersensitivity. On first exposure to an allergen, antigen-presenting cells (*APC*) phagocytose, degrade, and subsequently present allergen fragments to T cells. T cells help B cells make antibodies of the immunoglobulin E (*IgE*) isotype. Upon second exposure to the allergen, IgE-mediated cross-linking triggers mast cells or basophils to release vasoactive mediators, resulting in tissue damage and anaphylactic response. *TCR*, T-cell receptor; T_H2, T helper 2 cell; *PAF*, platelet activating factor.

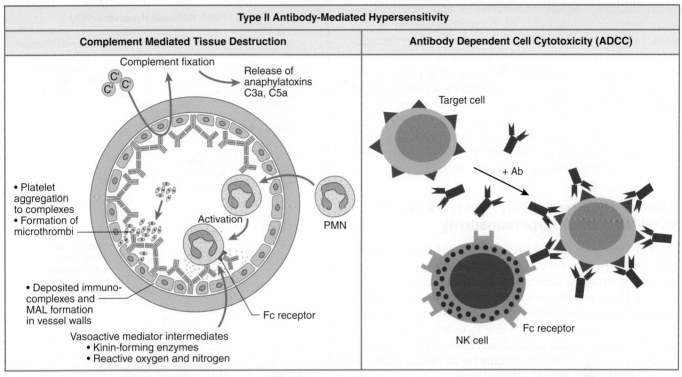

Figure 7-5. Type II antibody-mediated hypersensitivity. Type II reactions can occur through antibody recognition of surface antigen and subsequent deposition of complement; lysis is mediated through assembly of the complement membrane attack complex (*left*). Alternatively, bound antibodies are cross-linked through Fc receptors on polymorphonuclear (*PMN*) cells or natural killer cells, leading to antibody-dependent release of cytotoxic granules (*right*).

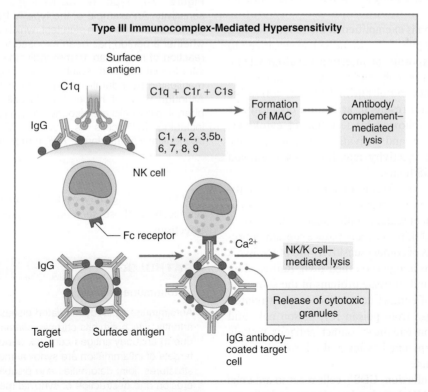

Figure 7-6. Type III immunocomplex-mediated hypersensitivity. Deposition of immunocomplexes in vascular beds results in platelet aggregation, complement fixation, and subsequent polymorphonuclear infiltration. Released factors from activated polymorphonuclear cells cause significant tissue damage and pathology, mediated by formation of the complement cascade membrane attack complex (*MAC*) or via lysis by released cytotoxic granules. *NK*, natural killer; *Ig*, immunoglobulin.

activation (C3a, C5a) are chemotaxins for acute inflammatory cells, resulting in infiltration by polymorphonuclear (*PMN*) cells. Lysosomal enzymes are released that result in tissue injury. Platelet aggregation occurs, resulting in microthrombus formation in the vasculature. This type of hypersensitivity was classically characterized as the **Arthus reaction**, identified by a high degree of PMN infiltrate, vasoactive amine release, and erythema and edema in response to intradermal injection of antigen. Type III reactions and accompanying inflammatory injury are seen in diseases such as rheumatoid arthritis, systemic lupus erythematosus, and postinfectious arthritis.

Type IV Hypersensitivity: Delayed-Type (Cell-Mediated) Hypersensitivity

Type IV hypersensitivity (also called **delayed-type hypersensitivity** [DTH]) involves T cell–antigen interactions that cause activation, cytokine secretion, and potential granuloma formation. This type of hypersensitivity requires sensitized lymphocytes that respond 24 to 48 hours after exposure to soluble antigen. DTH reactions may involve T_H cells (CD4+) or CTLs (CD8+ CTLs). Diseases such as tuberculosis, leprosy, and sarcoidosis as well as contact dermatitis are all clinical examples where tissue injury is primarily due to the vigorous immune response to released antigens rather than to the inciting pathogen itself. In these examples, sustained release of antigen and continued activation of sensitized T cells result in amplified tissue damage. Continual antigen release due to persisting infection may provoke excessive macrophage activation and granulomatous responses, leading to extended fibrosis and necrosis of tissue.

A classic DTH reaction is exemplified in the tuberculin skin test (Mantoux reaction) (Fig. 7-7). An individual sensitized to tuberculosis through exposure or infection develops CD4+ lymphocytes specific for mycobacterial antigens. Skin testing through intradermal injection of purified protein derivative (PPD) from mycobacteria results in activation of sensitized CD4+ T cells. This is followed by secretion of cytokines, which cause recruitment and activation of macrophages. The result is localized reactivity manifested by erythema and induration within 48 hours.

Another common form of DTH is allergic contact dermatitis, whose clinical presentation manifests as a pruritic, vesicular rash. In immunocompetent individuals, there is a requirement of repeated skin contact to induce subsequent reaction. Often, the agents of contact sensitivity are of too low a molecular mass to act as antigens on their own. Rather, they serve as haptens and bind to tissue proteins in the skin. The most common causes of contact dermatitis are plant-related antigens, including those from poison ivy, poison oak, and poison sumac. Common chemical contact antigens include formaldehyde, paraphenylenediamine and ethylenediamine, and potassium dichromate.

In instances of viral infection, CD8+ cells react to antigens presented via class I MHC molecules. Cytotoxic cells recognize the presented antigen and lyse the infected target.

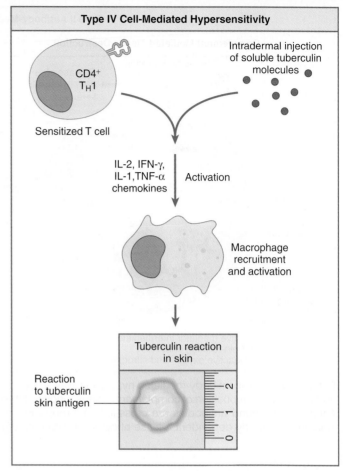

Figure 7-7. Type IV delayed-type (cell-mediated) hypersensitivity. An example of the type IV hypersensitive reaction is the Mantoux tuberculin skin test, which is performed to evaluate whether a person has been exposed to tuberculosis. A positive reaction of more than 10 mm indicates reactivity of T cells with intradermal injected soluble mycobacterial antigen. This is manifested by induration and inflammation at the injection site. Although CD4+ T helper 1 (T_H1) cells are shown here, other phenotypic T cells are also involved in type IV hypersensitivities, including T_H17 cells as well as CD8+ T lymphocytes. *IL*, interleukin; *IFN*, interferon; *TNF*, tumor necrosis factor.

Bystander killing may occur due to overaggressive responses, such as those seen in smallpox, measles, and herpes infections.

PATHOLOGY

Autoimmune Vasculitis

Autoimmune antibody-mediated diseases include rheumatoid arthritis, which exhibits chronic inflammation and vasculitis due to antibody-antigen complex deposition. The primary targets of inflammation are synovial membranes and articular structures; joint deformities and disability result from the erosion and destruction of synovial membranes and articular surfaces.

KEY POINTS ABOUT HYPERSENSITIVITY

- Hypersensitivity reactions fall into four major categories:
- Type I reactions represent allergic reactions (anaphylactic hypersensitivity).
- Type II reactions are antibody mediated and are characterized by cytotoxic events.
- Type III reactions are characterized by immunocomplex deposition and subsequent tissue destruction.
- Type IV reactions represent DTH responses that are controlled in large part by sensitized T cells reacting specifically the antigens.

phage activation and delayed-type hypersensitivity), whereas the T_H2 subset is active in antibody activation and allergic reactivity.

- Inappropriate responses may lead to immune-related pathologic conditions characterized by hypersensitivities.
- The different hypersensitivity reactions are grouped into four overlapping categories, loosely described as allergic or analphylactic (type I), antibody-mediated cytotoxic (type II), immune complex deposition (type III), or T-cell–mediated *DTH* (type IV).

Self-assessment questions can be accessed at www.StudentConsult.com.

KEY POINTS

- T cells are divided into helper or cytotoxic groups, with effector function regulating cellular immunity.
- Development of T_H lymphocytes is determined by the cytokine evironment present during intial antigen activation events.
- There are multiple subclasses of T_H lymphocytes. Regarding two major phenotypes, T_H1 cells control cellular immunity (macro-

Immunomodulation 8

CONTENTS
EVENTS INVOLVED IN T-LYMPHOCYTE ACTIVATION
TOLERANCE FOR SPECIFIC CONTROL OF IMMUNE RESPONSE
  Tolerance of Lymphocytes
  Elimination of Self-Reactive Lymphocytes
ETIOLOGY OF AUTOIMMUNE DISEASE
TRANSPLANTATION: CONTROL OF IMMUNE-MEDIATED REJECTION
  Graft Tissue
  Mechanisms of Allograft Rejection
  Immunosuppressive Agents to Prevent Allograft Rejection
  Graft-Versus-Host Disease
CANCER IMMUNOLOGY
IMMUNODEFICIENCY DISORDERS AS A PREDISPOSITION TO INFECTION
IMMUNOPROPHYLAXIS AGAINST INFECTIOUS AGENTS
IMMUNE THERAPY FOR CANCER

The regulation of immune function and overall immunohomeostasis is under control of multiple factors that include genetic and environmental components. Human leukocyte antigen (HLA) allotypes, antigen dose, and existing cytokine milieu can all influence responses to pathogenic agents. Age, nutritional status, and status of neuroendocrine molecules can also contribute in a complex relationship to cellular response. At one extreme, newborns are immunocompetent but immune-immature at birth; fetuses make immunoglobulin (Ig) M but not IgG until birth. Maternal IgG provides protection against bacterial agents through the first months of life. Extreme age is also a determinant. Immune senescence occurs in the elderly in whom good memory responses are available but poor primary (naive) response results in increased susceptibility to organisms and strains never before encountered. Both nutritional status and stress-related endocrine changes directly affect cell-mediated immune responses, most often leading to immunosuppression.

EVENTS INVOLVED IN T-LYMPHOCYTE ACTIVATION

Immune response must be regulated to be sufficient to protect the host without excessive or inappropriate responses (hypersensitivities) that may create disease. Activation of T cells requires two distinct signals: one through the T-cell response (TCR)/CD3 complex when interacting with antigen on the major histocompatability complex (MHC) and a second costimulatory signal. Receptor recognition of antigen mediates transcription of cytokine genes by a complex sequence of molecular events (Fig. 8-1). Absence of a secondary signal can lead to cellular inactivation. The professional presenting cells that provide costimulation are dendritic cells, macrophages, and B cells. Of these cell types, dendritic cells deliver the best costimulatory signals for activation of naive T cells. Presenting cells use the membrane molecules of the B7 family to deliver costimulatory signals through interaction with their ligand on T-cell membranes, CD28; two isoforms of the B7 family (B7.1 and B7.2; also called *CD80* and *CD86*) act to regulate cytokine responses. For example, CD80 binding upregulates T helper (T_H) 1 cytokines (interleukin [IL]-2, interferon [IFN]-γ), whereas CD86 binding upregulates T_H2 response (IL-4, IL-5, IL-6, and IL-10). Interactions with CD28 are critical in that binding to CD28 on the lymphocyte leads to IL-2 production, a major T-cell growth factor, whereas inhibition of binding leads to development of tolerance or even cellular energy (non-responsivenes). Interaction of the B7 molecules with other T-cell surface antigens, such as CD152 (CTLA-4), leads to supressive signals and tolerance/anergy, while induction of alternate signalling leads to memory cell formation.

KEY POINTS ABOUT EVENTS INVOLVED IN T-LYMPHOCYTE ACTIVATION

- Immune responses are regulated to allow sufficient response without development of hypersensitivity.
- Activation of T cells requires two distinct signals: one by TCR recognition of MHC-presented antigen and another through costimulatory molecule interactions.

TOLERANCE FOR SPECIFIC CONTROL OF IMMUNE RESPONSE

The major goal of the adaptive immune response is recognition of specific antigens, for which the specificity is controlled to allow recognition of foreign (nonself) antigens. The intensity and duration of response must be sufficient to protect against invading pathogens, with prompt and specific downregulation when foreign antigen is no longer present. *Immunologic tolerance* is the failure to mount an immune response to an antigen. Immunologic tolerance is not simply the failure to recognize an antigen but rather an active response to a particular epitope that allows specific unresponsiveness while the rest of the host response remains intact and capable of response.

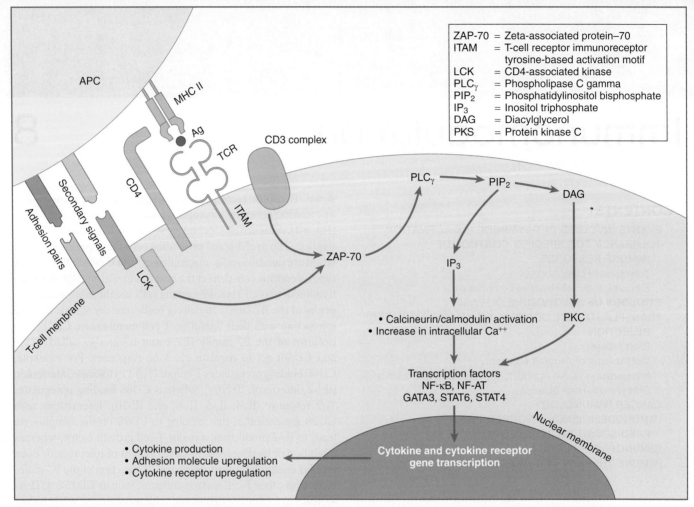

Figure 8-1. Receptor-mediated transcription of cytokine genes by a sequence of molecular events. T-cell receptors' recognition of antigen presented (*APC*) by major histocompatability complex (*MHC*) molecules triggers transduction of signals across the cell membrane through the CD3 complex, resulting in upregulation of T-cell activation genes (cytokines and cytokine receptors). *Ag*, antigen; *TCR*, T-cell response; *PKC*, protein kinase C.

Tolerance of Lymphocytes

Tolerance can be at the level of B or T lymphocytes, or both, and can be accomplished by a variety of mechanisms including apoptosis, anergy, antigen excess, and suppressor/regulatory T-cell formation (Fig. 8-2). T-regulatory cells (CD4+, CD25+, Foxp3+) also play a major role in modification of immune response, especially to limit development of autoimmune reactivity. Only lymphocytes bearing antigen receptors can be tolerized. Tolerance can be induced early during adaptive cell development, or peripherally during response maturation. Two mechanisms for tolerance include **apoptosis**, or programmed cell deletion, and **anergy**, a state of clonal inactivation due to lack of secondary signals.

Elimination of Self-Reactive Lymphocytes

Induction of tolerance in immature lymphocytes is critical for elimination of self-reactive cells. For T lymphocytes, negative selection via apoptosis in the thymus eliminates the majority of cells that bear high reactivity to self-peptides. Mature cells leaving the thymus may also be tolerized, subject to antigenic factors that affect immunogenicity. Factors that induce tolerance include extremely high- or low-dose antigen, weak MHC binding, and route of administration (e.g., oral antigens are well tolerated). Peripheral tolerance occurs primarily via anergic reactions, where costimulatory signals provided by the B7 (molecules called CD80/CD86) on the surface of the presenting cell are absent. Finally, tolerance may also occur in immunologically privileged sites, such as the eye, brain, and testes, usually due to local secretion of immunosuppressive factors (e.g., tumor growth factor-β and/or IL-10) that downregulate response. The developing fetus in the uterus is a prime example of a privileged site in which recognition to foreign antigen is suppressed.

Immature B cells in the bone marrow undergo apoptosis upon recognition of self-antigens by activation-induced cell death mechanisms. Alternatively, the B cell may undergo receptor editing to change the binding specificity of the surface immunoglobulin, thus rendering the cell no longer capable of self-reactivity. Once in the peripheral tissue, B cells may undergo an anergic response dependent upon the level of specific antigen. Low-dose soluble monomeric antigens

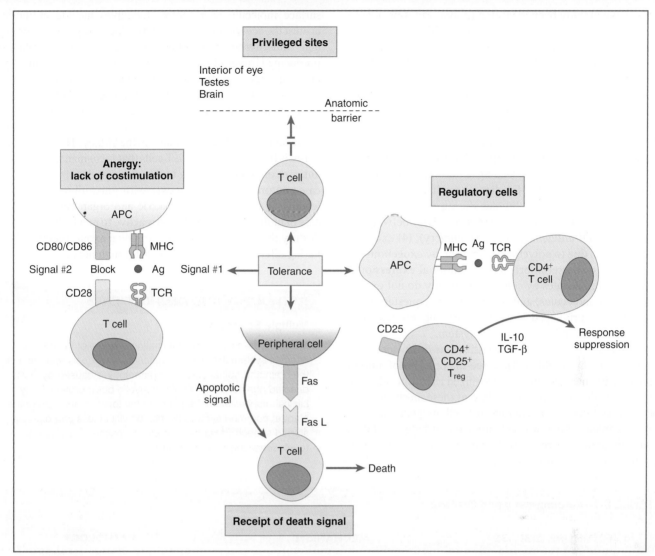

Figure 8-2. Tolerance can occur by mechanisms including apoptosis of reactive cells, development of anergic response to antigen through loss of secondary signals, regulation of response as a result of antigen excess, and active suppression by regulatory T cells (T_{reg}). *APC*, antigen-presenting cell; *MHC*, major histocompatability complex; *Ag*, antigen; *TCR*, T-cell receptor, *IL*, interleukin; *TGF*, tumor growth factor.

do not permit receptor cross-linking on the surface of the B cell, sending signals to clonally inactivate the B cell. An excessively high antigen dose also can result in anergic response owing to overwhelming recognition in the absence of sufficient T-cell costimulation. Finally, self-reactive B cells that escape elimination or induction of anergy may be rendered incapable of activation as a result of the lack of T cells available to help initiate development of autoimmune response.

ANATOMY AND EMBRYOLOGY

Apoptosis during Vascular Development

The development of vasculature to serve as a nutrient and waste pipeline is fundamental for organ development embryogenesis. Although the molecular determinants of capillary regression are as yet unresolved, apoptosis as a mechanism for cell death is critical in fetal development for regionalization of mesoderm in embryonic growth.

KEY POINTS ABOUT TOLERANCE FOR SPECIFIC CONTROL OF IMMUNE RESPONSE

- Immunologic tolerance represents specific unresponsiveness to antigen while remaining responses are intact.
- Factors influencing the nature, intensity, and duration of immune function include age, neuroendocrine hormone levels, HLA allotypes, antigen dose, and cytokine milieu.
- Tolerance can be at the level of B or T lymphocytes, or both, and can be accomplished by mechanisms including apoptosis, anergy, antigen excess, and suppressor/regulatory T-cell function.

ETIOLOGY OF AUTOIMMUNE DISEASE

Failure of tolerance can lead to development of autoimmunity. It is clear that both environmental and genetic components are risk factors for autoimmune disease. Many genetic factors have already been identified (polymorphisms of cytokine genes, such as tumor necrosis factor [TNF], or their receptors; apoptosis genes; complement component deficiencies), and others are likely to be revealed soon. In addition, infections or exogenous agents that cause physical damage are likely to play important roles. The mechanisms underlying all autoimmune diseases are not fully elucidated; however, genetic polymorphisms of MHC class II genes (alleles of HLA-DR and/or HLA-DQ) are strongly associated with increased susceptibility to autoimmune diseases. Possible mechanisms for a loss of tolerance leading to autoimmune reactions include: (1) a lack of Fas-Fas ligand–mediated deletion of autoreactive T cells in the thymus during development, (2) loss of T-regulatory function, (3) cross-reactivity between exogenous and self-antigens (molecular mimicry), (4) excessive B-cell function due to polyclonal activation by exogenous factors (of viral or bacterial origin), (5) abnormal expression of MHC class II molecules by cells that normally do not express these surface molecules, and (6) release of sequestered self-antigens from privileged sites, thus priming for responses not previously seen by the immune system.

Autoimmune diseases can be classified as organ-specific or systemic in nature (Table 8-1). Three major types of autoimmune reaction mechanisms are recognized as causing different autoimmune disorders (Fig. 8-3). Two of these mechanisms involve autoantibodies directed against self-antigens; for both, classic complement pathway activation exacerbates local damage and inflammatory response. In the first case, autoantibodies may be directed against a specific self-component, such as a surface molecule or receptor. Examples include antibodies against the acetylcholine receptor producing myasthenia gravis, and anti–thyroid-stimulating hormone receptor antibodies producing Graves disease. Autoantibodies may also bind with antigens present in the blood, forming antigen-antibody (immune) complexes that later deposit in organs, thus inciting an inflammatory response. An example is seen in lupus λ-related glomerulonephritis, in which complexes of anti-DNA antibodies and free DNA accumulate in the kidney. The third mechanism is that of autoreactive T cells that recognize targeted self-antigens on organs, leading to direct damage to tissue. In many cases, autoreactive T cells coexist with autoantibody responses, leading to exacerbation of disease and organ damage. In the case of multiple sclerosis, T cells reactive to myelin basic protein destroy the protective layer surrounding axons, thereby eliminating effective transfer of signals through nerves.

PATHOLOGY AND NEUROSCIENCE

Multiple Sclerosis

Multiple sclerosis is a demyelinating disease of the central nervous system that leads to patches of sclerosis (plaques) in the brain and spinal cord. The pathology is caused by TCRs directed against antigens of the myelin basic protein. Early clinical manifestations depend on the location and size of the plaque; common symptoms include visual loss and diplopia, slurred speech (dysarthria), muscle weakness, and abnormal touch sensation (paresthesia).

TABLE 8-1. Autoimmunity and Disease

AUTOIMMUNE DISEASE	MECHANISM	PATHOLOGY
Autoimmune hemolytic anemia	Autoantibodies to RBC antigens	Lysis of RBCs and anemia
Autoimmune thrombocytopenia purpura	Autoantibodies to platelet integrin	Bleeding, abnormal platelet function
Myasthenia gravis	Autoantibodies to acetylcholine receptor in neuromuscular junction	Blockage of neuromuscular junction transmission and muscle weakness
Graves disease	Autoantibodies to receptor for thyroid-stimulating hormone	Stimulation of increased release of thyroid hormone (hyperthyroidism)
Hashimoto thyroiditis	Autoantibodies and autoreactive T cells to thyroglobulin and thyroid microsomal antigens	Destruction of thyroid gland (hypothyroidism)
Type 1 diabetes	Autoantibodies and autoreactive T cells to pancreatic islet cells	Destruction of islet cells and failure of insulin production
Goodpasture syndrome	Autoantibodies to type IV collagen	Glomerulonephritis
Rheumatic fever	Autoantibodies to cardiac myosin (cross-reactive to streptococcal cell wall component)	Myocarditis
Pemphigus vulgaris	Autoantibodies to epidermal components (cadherin, desmoglein)	Acantholytic dermatosis, skin blistering
Multiple sclerosis	T-cell response against myelin basic protein	Demyelination, marked by patches of hardened tissue in the brain or the spinal cord; partial or complete paralysis and jerking muscle tremor
Systemic lupus erythematosus	Circulating immunocomplexes deposited in skin, kidneys, etc., formed by autoantibodies to nuclear antigens (antinuclear antibodies), including anti-DNA	Glomerulitis, arthritis, vasculitis, skin rash

Continued

TABLE 8-1. Autoimmunity and Disease—cont'd

AUTOIMMUNE DISEASE	MECHANISM	PATHOLOGY
Rheumatoid arthritis	Autoantibodies to immunoglobulin G (rheumatoid factors); deposition of immunocomplexes in synovium of joints and elsewhere; infiltrating autoreactive T cells in synovium	Joint inflammation, destruction of cartilage and bone
Celiac disease	Antibodies made to gliadin (gluten), cross-reactive to tissue transglutaminase	Gluten-sensitive enteropathy, villous destruction and gastrointestinal manifestations
Scleroderma (systemic sclerosis)	Antibodies to topoisomerases, polymerases, and fibrillarin	Skin-related fibrosis, damage to related arteries
Ankylosing spondylitis	CD4+ cells, possible activity to self-antigens (arthritogenic peptides, molecular mimicry, or aberrant forms of B27)	Rheumatic disease of joints and spine
Sjögren syndrome	CD4+ cells, possible activity to self-antigens (M3 muscarinic acetylcholine receptor)	Lymphocytic-mediated destruction of lachrymal and salivary glands

RBC, red blood cell.

KEY POINTS ABOUT THE ETIOLOGY OF AUTOIMMUNE DISEASE

- Autoimmunity represents a failure of effective tolerance to self-antigens.
- Genetic and environmental factors play a role in the etiology of disease.
- Mechanisms of disease include autoantibodies that are directed against specific self-components, deposition of circulating antibody-antigen complexes, and deleterious responses by autoreactive T cells.

TRANSPLANTATION: CONTROL OF IMMUNE-MEDIATED REJECTION

Graft Tissue

Understanding of the mechanisms involved in immunoregulation has allowed increased success rates of tissue transplantation in the clinic. It has long been known that tissue transplanted from an individual to another location within the same individual (**autografts**) is well tolerated. Similarly, organs transplanted between identical twins or animals from an inbred strain of the same species (**syngeneic grafts** or **isografts**) have a high degree of success. By contrast, grafts transplanted between individuals of different species (**xenografts** or **heterografts**) are rarely successful without a high degree of immunosuppression. Indeed, grafts made between genetically different individuals of the same species (**allografts** or **homografts**) are tolerated only when MHC antigens are matched. The common cause of graft rejection is due to T-cell recognition of foreign HLA antigens on the donor tissue. Histocompatibility testing is used to identify HLA class I and class II antigens, as a measure to maximize the number of alleles conserved between donor and recipient individuals. Alloreactivity may therefore be defined as the stimulation of host TCRs against foreign HLA antigens. However, even with a high degree of HLA similarity, minor (non-HLA) alloantigens may induce rejection of allografts.

Mechanisms of Allograft Rejection

The mechanisms for allograft rejection are defined by the speed at which the tissue is attacked by the host (Fig. 8-4). Hyperacute rejection occurs within minutes after tissue transplantation and is mediated by preformed antibodies present and directed against HLA or blood group antigens. Antibodies bound to vasculature activate complement components, causing immediate obstruction and destruction of donor blood vessels. This is in contrast to acute rejection, in which T cells responding to donor tissue antigens are amplified and activated. Acute rejection occurs approximately 10 to 30 days after tissue implantation. Even if tissue is matched for major histocompatibility antigens, the minor HLA components may be enough to eventually cause rejection of tissue. Chronic rejection, therefore, manifests as reactions occurring months to years after transplantation, involving both T cells and antibodies directed against minor antigens; infiltration of CD4+ and CD8+ T cells causes release of cytokines and direct lysis of target tissue, eventually resulting in graft rejection.

Immunosuppressive Agents to Prevent Allograft Rejection

At present there is no clinical protocol to induce complete tolerance to allografts. Therefore, all patients require daily treatment with immunosuppressive agents. All immunosuppressive agents in clinical use have drawbacks relating either to toxicity and side effects or to failure to provide sufficient immunosuppression. On the one hand, excessive immunosuppression can lead to development of opportunistic infections and neoplasia. On the other hand, inadequate immunosuppression allows the

Immunomodulation

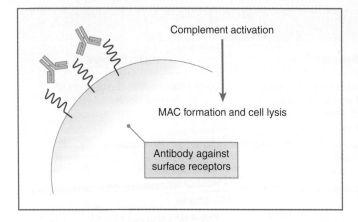

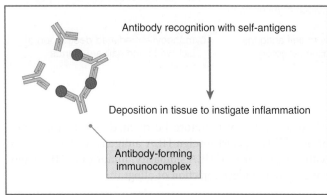

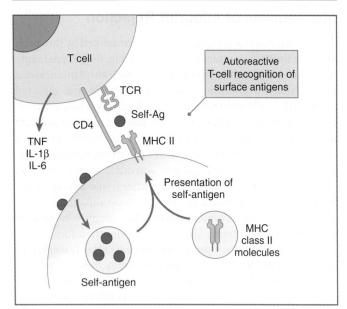

Figure 8-3. Autoimmunity and disease. *MAC*, membrane attack complex; *TCR*, T-cell response; *Ag*, antigen; *MHC*, major histocompatability complex; *TNF*, tumor necrosis factor; *IL*, interleukin.

recipient to mount an immune response, causing allograft rejection. Table 8-2 lists immunosuppressive agents used in clinical transplantation.

Graft-Versus-Host Disease

One consequence of increased graft acceptance due to immunosuppressive drugs is graft-versus-host disease (GVHD). A "reverse rejection" occurs whereby healthy donor lymphocytes recognize the patient's tissues as foreign. In GVHD, persistent lymphocytes proliferate and react against the host antigens; over time, this expanded population targets the skin, bowel, and liver, as well as other tissues.

PATHOLOGY

GVHD

T cells expanded from within an allogeneic donor organ transplant (e.g., bone marrow transplant) react against the recipient's antigens, attacking cells and tissues. Disease extends to affect the skin, gastrointestinal tract, and liver, with symptoms including rash, fever, diarrhea, liver dysfunction, abdominal pain, and anorexia.

CANCER IMMUNOLOGY

Immune surveillance is a natural physiologic function to allow recognition and destruction of transformed cells before they grow into tumors, and to kill tumors after they are formed. The exquisite recognition of antigens based on foreignness, high molecular weight, complexity, and degradability must be balanced with mechanisms of tolerance induction to limit tumor expansion. In addition, tumor cells have the capacity to limit responsiveness through various mechanisms, including downregulation of MHC expression, loss of expression of antigenicity, failure to express costimulatory molecules to activate cellular responses, and via direct production of activation cytokines.

The principal mechanism of tumor immunity is through function of CD8+ cells, with assistance from CD4+ T_H1 cell phenotypes. However, natural killer (NK) cells also play a major role, in that they kill tumorigenic cells without direct need for sensitization or MHC restriction; they can directly lyse targets based on immunoglobulin recognition of antigen accompanied by local secretion of TNF-α. B cells are also a major defense component; however, their function is poorly understood. It is known that IgM and IgG promote protection, especially in the presence of complement factors.

KEY POINTS ABOUT MECHANISMS IN CANCER IMMUNITY

- T-cell function via cytolysis and apoptosis is critical for rejection of viral-induced tumors.
- NK cells function by direct cytolysis, apoptosis, and antibody-dependent cell cytotoxicity to limit cells that poorly express MHC antigens.
- Lymphokine-activated killer cells also function via direct cytolytic function.
- Activated macrophages can enhance protection by phagocytosis and cytolysis of tumorigenic or neoplastic cells.
- Cytokines can cause direct effects to kill neoplastic cells as well as recruit other inflammatory cells to local sites.
- The role of B cells in limiting growth of tumors is not well understood, but may function via complement and osponization mechanisms.

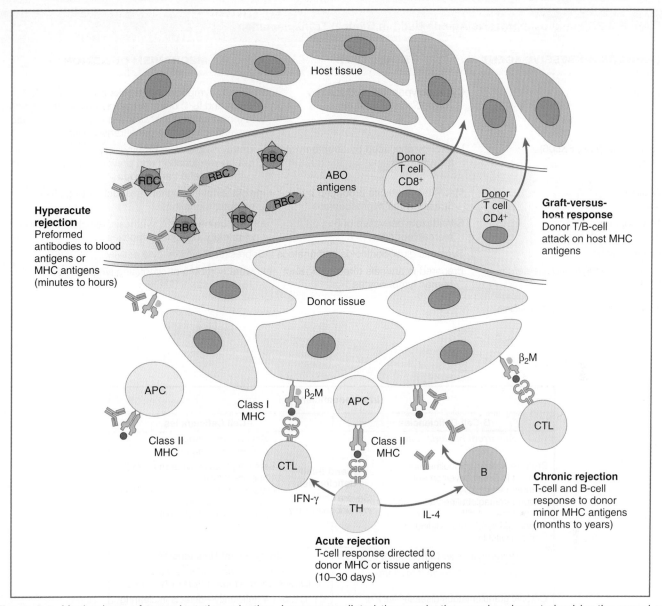

Figure 8-4. Mechanisms of transplantation rejection. Immune-mediated tissue rejection can be characterized by the speed of response to donor tissue and the effector mechanisms involved. *RBC*, red blood cell; *MHC*, major histocompatability complex; *APC*, antigen-presenting cell; β_2M; β_2-microglobulin; *CTL*, cytotoxic T lymphocyte; *TH*, T helper cell.

●●● IMMUNODEFICIENCY DISORDERS AS A PREDISPOSITION TO INFECTION

Immunodeficiency disorders are a diverse group of illnesses that result from one or more abnormalities of the immune system (Fig. 8-5). The abnormalities can involve absence or malfunction of blood cells (lymphocytes, granulocytes, monocytes) or soluble molecules (antibodies, complement components) and can result from an inherited genetic trait (primary) or from an unrelated illness or treatment (secondary). The principal manifestation of immunodeficiency is an increased susceptibility to infection as documented by increased frequency or severity of infection, prolonged duration of infection with development of an unexpected complication or unusual manifestation, or infection with organisms of low pathogenicity.

KEY POINTS ABOUT IMMUNODEFICIENCY DISORDERS AS A PREDISPOSITION TO INFECTION

- Immunodeficiencies lead to disease susceptibility dependent upon loss of immune function.
- B-cell deficiency is marked by recurrent infections with encapsulated bacteria.
- T-cell deficiency manifests as recurrent viral, fungal, or protozoal infections.
- Phagocytic deficiency with the associated inability to engulf and destroy pathogens usually appears with recurrent bacterial infections.
- Complement disorders demonstrate defects in activation patterns of the classical, alternative, and/or lectin-binding pathways, which augment adaptive host defense mechanisms.

TABLE 8-2. Immunosuppressive Agents Used in Clinical Transplantation

IMMUNOSUPPRESSIVE AGENT	STRUCTURE	MECHANISM OF ACTION
Cyclosporine	Cyclic polypeptide produced by *Tolypocladium inflatum*	Inhibits T_H lymphocytes; blocks production of IL-2, IFN-γ, and IL-4; binds to cytoplasmic immunophilin to inhibit calcineurin, affecting nuclear factors and preventing cytokine gene transcription
Tacrolimus (TCL; FK506)	Macrolide produced by *Streptomyces tsukubaensis*	Forms a complex with FK-binding protein; also blocks calcineurin function and prevents IL-2 (and other cytokine) gene transcription
Sirolimus	Macrolide antibiotic produced by *Streptomyces hygroscopicus*	Inhibits cytokine signal transduction (e.g., IL-2 receptor signaling)
Azathioprine	S-Imidazole derivative of 6-mercaptopurine	Inhibits de novo DNA synthesis, affecting initiation of primary immune responses
Corticosteroids	Synthetic glucocorticoids and analogs	Depresses T-lymphocyte activity
Polyclonal antilymphocyte serum	Prepared in animals directed against human lymphocytes	Lowers numbers of circulating lymphocytes

TH, T helper; *IL*, interleukin; *IFN*, interferon.

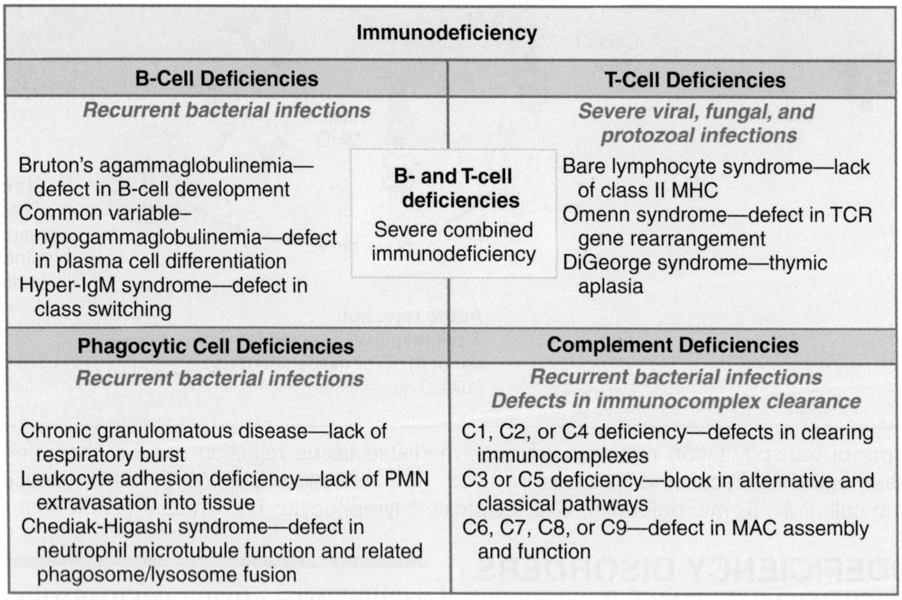

Figure 8-5. Primary immunodeficiencies. Manifestation of immunodeficiency depends on the etiology of response. B-cell deficiency is marked by recurrent infections with encapsulated bacteria. T-cell deficiency manifests as recurrent viral, fungal, or protozoal infections. Phagocytic deficiency with associated inability to engulf and destroy pathogens usually manifests with recurrent bacterial infections. Complement disorders demonstrate defects in activation patterns of the classical, alternative, and/or lectin-binding pathways, which would otherwise naturally augment adaptive host defense mechanisms. *IgM*, immunoglobulin M; *PMN*, polymorphonuclear cell; *MHC*, major histocompatability complex; *TCR*, T-cell response; *MAC*, major attack complex.

●●● IMMUNOPROPHYLAXIS AGAINST INFECTIOUS AGENTS

The active process of exposure to infectious agents or their antigenic components (vaccination or immunoprophylaxis) has been one of the most important medical advances that directly affect longevity and quality of life. Immunization may be characterized as active, whereby exposure to antigen generates protective immunity of the host. Immunization may also be passive, in which administered humoral (antibody) or cellular factors confer host resistance to disease. Active immunization provides long-lasting immunity, whereas passive administration of factors leads to fast-acting but temporary immunity to eminent or ongoing exposure.

TABLE 8-3. Schedule for Active Immunization of Children, Adults, and Special Populations

AGE	VACCINE
Birth	Hep B
1–2 months	Hep B
2 months	DTP, Hib, IPV
4 months	DTP, Hib, IPV, Rv
6 months	Hep B, DTP, Hib, IPV, Rv
12–15 months	OPV, MMR, varicella vaccine for susceptible children
1–18 years	HBV
4–6 years	DTP, OPV, MMR
11–12 years	Hep B, MMR, varicella
11–18 years	HPV
25–64 years	Measles, rubella
>65 years	Influenza, pneumococcal disease
Special Populations	
Health care workers	Hep A, Hep B, influenza
Military personnel	Adenovirus, encephalitis viruses, yellow fever, *Bacillus anthrasis* (anthrax)
Animal caretakers	Rabies, *Francisella tularensis* (tularemia), *Yersinia pestis* (plague), *Neisseria* (meningococcal)
Travelers at risk	*Vibrio cholerae* (cholera), *Salmonella typhi* (typhoid), yellow fever, Japanese encephalitis virus, Hep A

Hep, hepatitis; *DTP,* diphtheria and tetanus toxoids and acellular pertussis; *Hib, Haemophilus influenzae* type b; *IPV,* inactivated poliovirus; *Rv,* rotavirus; *OPV,* oral poliovirus vaccine; *MMR,* measles, mumps, rubella; *HPV,* human papilloma virus.

Vaccination schedules require understanding of the immune status of the developing individual (Table 8-3). The fetus begins to produce IgM at 6 months' gestation; most IgG at birth is transplacental and maternal in origin. Infants do not respond well to polysaccharide antigens prior to 2 years of age (Fig. 8-6). Antigenicity of small molecules (haptens) improves when they are conjugated to protein. Vaccines also contain **adjuvants,** which are excipients included in the vaccine formulation to stimulate immune function toward production of protective responses against pathogenic antigens.

KEY POINTS ABOUT IMMUNOPROPHYLAXIS AGAINST INFECTIOUS AGENTS

- Immunization may be active or passive.
- Active immunity is the development of antibodies in response to antigenic stimulation.
- Passive immunity is transferred protection through preformed antibodies produced by another individual.

IMMUNE THERAPY FOR CANCER

Immune activation is critical for control of tumor expression, with a need to augment existing immune surveillance processes to augment protection against development of cancerous tissue. Novel therapeutic approaches seek to augment natural immune function by eliciting anti-tumor response, expanding

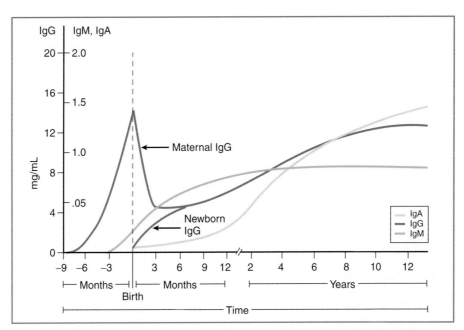

Figure 8-6. Developmental progression of antibody isotypes. Maternal antibodies wane through the first year as the child produces antibodies of his or her own. *Ig,* immunoglobulin.

tumor-reactive T cells, and suppressing T-regulatory cells that affect immune tolerance to self-antigens. Novel adjuvants, such as α-galactosylceramide, which activates NKT cells, toll-like (TLR) agonists that stimulate presenting cells, and monoclonal antibodies, such as those that negatively regulate immune-activating receptors, are part of a new wave of therapeutic anticancer agents. Table 8-4 lists immune therapies that have been successful against cancer modalities.

TABLE 8-4. Immune-Based Therapeutics Against Cancer

THERAPY	INDICATION
Monoclonal antibodies	Non-Hodgkin lymphoma, colorectal and lung cancer
Adjuvants (BCG)	Bladder cancer
Cytokines (IFN-α, IL-2, TNF-α)	Melanoma, sarcoma
Bone marrow transplant	Hematogenous malignancy

BCG, Bacille Calmette-Guérin; INF, interferon; IL, interleukin; TNF, tumor necrosis factor.

KEY POINTS

- T lymphocytes are activated through (1) specific interaction of their TCR with antigen presented by MHC molecules, and (2) engagement of costimulatory molecules present on the surface of the T cell and the APC. Absence of costimulatory interaction with the APC will lead to T-cell anergy.

- Tolerance may be at the level of the T cell or the B cell, induced by apoptosis, anergy, antigen excess, or T regulatory cell function. Immunologic tolerance represents specific unresponsiveness to antigen while other responses remain intact.

- A loss of effective tolerance to self often culminates in manifestation of autoimmune-related pathology.

- Mechanisms of transplantation rejection are defined by the presence of preformed antibody mediators and development of T-cell adaptive reactivity. GVHD reactions occur when immune cells within the donor tissue respond to host factors.

- Immune deficiencies can be subdivided into groups according to deficiencies in innate complement or phagocytic cell function or deficits in diferent classes of lymphocytic function.

- Vaccination allows development of protective immunity before encounter with antigen. Immunization may be active or passive.

Self-assessment questions can be accessed at www.StudentConsult.com.

Immunoassays 9

CONTENTS

INTERACTIONS BETWEEN POLYCLONAL ANTIBODIES AND MULTIVALENT ANTIGENS
ANTIBODY-ANTIGEN INTERACTIONS: THE BASIS OF QUANTITATIVE AND QUALITATIVE ASSAYS
ANTIBODY-ANTIGEN PRECIPITATION REACTIONS IN GELS
   Double Diffusion Assay (Ouchterlony Assay)
   Radial Immunodiffusion Assay
MONOCLONAL AND POLYCLONAL ANTIBODY SPECIFICITY FOR ANTIGEN DETECTION
   Western Blot (Immunoblotting) Detection of Membrane-Bound Antigens
   Enzyme-Linked Immunosorbent Assay
   Immunohistochemical Methods
   Fluorescence-Activated Cell Sorting
ASSESSMENT OF IMMUNE FUNCTION
   General Inflammation Assessment
   Complement Fixation Test
   Lymphocyte Function Assays
   Microarrays to Assess Gene Expression
   Animal Models to Study Human Disease
   ABO Blood Groups and Rh Incompatibility

Immunoassays and experimental systems are designed to assist in the assessment of pathogenic responses and are instrumental in our understanding of the pathophysiology of diseases. The exquisite specificity of the antibody allows its use as an efficient analytical tool in biomedical research and diagnostic investigation. Cellular assays allow determination of lymphocyte function, and gene expression arrays allow in silico identification of molecular mechanisms that form the basis of disease.

INTERACTIONS BETWEEN POLYCLONAL ANTIBODIES AND MULTIVALENT ANTIGENS

The use of antibodies has expanded from simple diagnostics to elucidation of gene function, localization of gene products, and rapid screening of biologic effectors for drug discovery and testing. The interactions between antigen and antibody form the basis of many quantitative and qualitative diagnostic assays that involve noncovalent binding of antigenic determinants (**epitopes**) to the variable regions of both the heavy and the light immunoglobulin chains. These interactions are analogous to those observed in enzyme-substrate interactions and can be defined similarly through physical laws of mass action. Specifically, the reversible interaction between antibody and an antigen can be represented by the equation:

$$K = \frac{k_1}{k_{-1}} = \frac{[AbAg]}{[Ab][Ag]}$$

where K represents the equilibrium constant, k_1 is the association constant, k_{-1} is the dissociation constant, Ab represents the free antibody concentration, Ag represents the free antigen concentration, and AbAg represents complexed antibody with antigen.

Although each individual antibody is highly specific for antigenic structures, the population of serum antibodies is typically polyclonal in nature. This means that complex antigens give rise to multiple antibodies that react with different determinants inherent within the antigen. Because serum (immunoglobulin G) antibodies are bivalent in their reactions with antigen, and because antigens are often multivalent in their interactions with antibodies, there exists a natural capacity for polyclonal antibodies to cross-link antigens.

BIOCHEMISTRY

Affinity Constants

Affinity constants determining antibody-antigen interactions may be determined by analysis of the equilibrium association or dissociation constants. The equilibrium association constant and valence of an antibody can be found using the Scatchard equation, $r/c = K(n - r)$, where r is the moles of bound ligand per mole of antibody at equilibrium, c is the free ligand concentration at equilibrium, K is the equilibrium association constant, and n is the number of antigen-binding sites per antibody molecule.

ANTIBODY-ANTIGEN INTERACTIONS: THE BASIS OF QUANTITATIVE AND QUALITATIVE ASSAYS

Experimentally, if a known concentration of antibody is mixed with increasing amounts of specific antigen, then cross-linked antibody-antigen complexes begin to precipitate from the solution. Because of the nature of the number of potential binding sites, an equivalence zone of maximal precipitation occurs that is characteristic of the interaction. If greater amounts of antigen are further added, the complexes will begin to dissolve and return to solution (Fig. 9-1). As discussed in Chapter 3, the antigen-antibody reactions are defined by **noncovalent forces, affinity,** and **avidity.**

The antibody titer reflects the measured amount of antibodies (in the blood), described as a relative dilution factor against a particular type of antigen (tissue, cell, or substance). Antibodies directed against epitopes on multideterminant antigens lead to agglutination reactions. The most widely described agglutination reaction is the **Coombs test**, originally used to detect autoantibodies against host red blood cells (RBCs) (Fig. 9-2). Many diseases and drugs (e.g., quinidine, methyldopa, and procainamide) can lead to production of these antibodies, with repercussions of RBC destruction and resultant anemia. In direct agglutination tests, the agglutinating antibodies directly react with antigens on the surface of bacterial cells (e.g., *Brucella* and tularemia febrile agglutinin tests) or erythrocytes (direct hemagglutination test) to form visible clumps of particles. In passive hemagglutination tests, agglutinating antibodies react with antigens coupled to carrier particles such as erythrocytes (indirect or passive hemagglutination test) or gelatinous beads to form visible aggregates.

●●● ANTIBODY-ANTIGEN PRECIPITATION REACTIONS IN GELS

Often, precipitation reactions are used for analysis in situations in which only qualitative detection is required. Reactions can be performed in a gel matrix to limit the rate of diffusion of reactants and hold the precipitate so that it is effectively immobilized for visualization. Several methods are widely used in medicine for analysis of hormones, enzymes, toxins, and immune system products.

Double Diffusion Assay (Ouchterlony Assay)

The double diffusion assay, also known as the Ouchterlony assay, was developed in the 1950s and is still in widespread use. The method is extremely simple and allows determination of identity (relatedness) between reagents. The most widespread

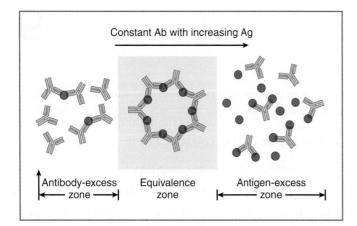
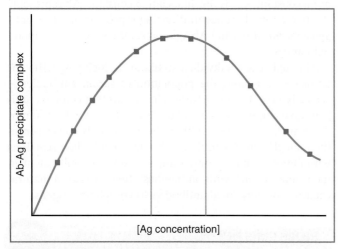

Figure 9-1. Physical interactions dictate antibody (*Ab*)-antigen (*Ag*) interactions.

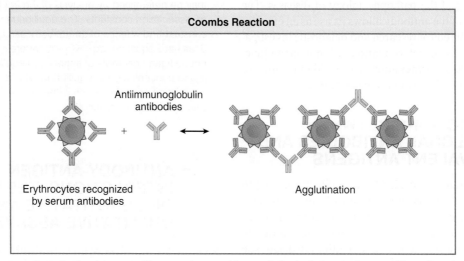

Figure 9-2. Coombs reaction. Antibody recognition can be used to agglutinate large antigens, either directly or indirectly through cross-linking with antiimmunoglobulin antibodies. Pictured is an indirect Coombs reaction in which autoantibodies recognizing determinants on erythrocytes are agglutinated using reactive anti-immunoglobulin antibodies.

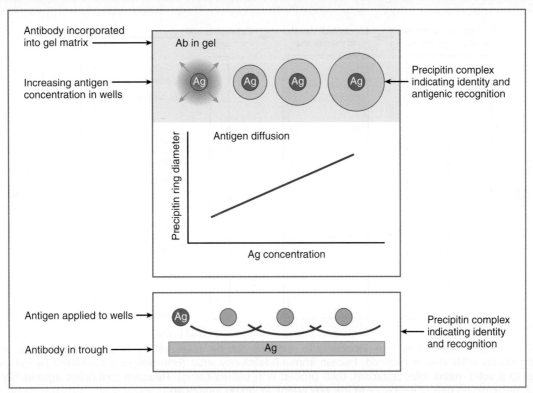

Figure 9-3. Double diffusion assay. The basis for the Ouchterlony assay (*top*) is the precipitin reaction which occurs when applied antigen (*Ag*) diffuses through a gel matrix containing reactive antibodies (*Ab*). The diameter of the precipitin ring formed can be quantitated by comparison to known standard antigen concentrations. The radial diffusion assay (*bottom*) allows qualitative determination as antibody and antigen diffuse toward each other through the assay matrix.

use of the Ouchterlony technique is for comparison of antigens (Fig. 9-3). It is used in forensic medicine and in a variety of diagnostic assays.

Radial Immunodiffusion Assay

A variation of the double diffusion assay allows quantitative measurement of antigen. The radial immunodiffusion assay (see Fig. 9-3) incorporates antibody directly into the assay gel matrix. Antigen is applied to wells and allowed to diffuse through the gel. A precipitin forms as the antigen and antibody interact; the amount of antigen applied to the well is proportional to the diameter of the precipitin. Comparison of unknown samples to known antigen concentrations allows quantitative analysis.

●●● MONOCLONAL AND POLYCLONAL ANTIBODY SPECIFICITY FOR ANTIGEN DETECTION

Antibodies in serum are by nature polyclonal, with potential to recognize more than 10^{14} different antigenic epitopes. Exposure to foreign molecules (nonself), either by infection or by immunization, will increase the relative balance and titer of polyclonal antibodies toward a specific antigen. The ability to recognize multiple epitopes is paramount to achieving protection against pathogens. However, it is often critical to have a large pool of antibodies with single epitope specificity that can be used for diagnostic and therapeutic purposes. Monoclonal antibodies are made in large quantities by single hybrid clones. They are especially well suited as analytical reagents for sensitive and specific detection of defined molecular epitopes, and are commonly used for phenotypic identification.

Western Blot (Immunoblotting) Detection of Membrane-Bound Antigens

A mixture of proteins (e.g., antigens derived from pathogenic organisms) may be separated by electrophoresis through a gel matrix and then transferred onto a solid medium, such as nitrocellulose, that binds proteins tightly. Antibodies with a reporter enzyme covalently attached are incubated with proteins bound to the nitrocellulose. Substrate for the enzyme is added, which develops color when enzyme is present. The resulting colored line demonstrates that the antigen was present in the initial population of mixed proteins. This type of test is clinically useful for detection of patient antibodies reactive to viral antigens, such as the *gag*, *pol*, or *env* gene products produced by the human immunodeficiency virus (Fig. 9-4).

Enzyme-Linked Immunosorbent Assay

The two general approaches to diagnosing disease by immunoassay include methods that directly test for specific antigens or indirectly test for the presence of antigens by looking for

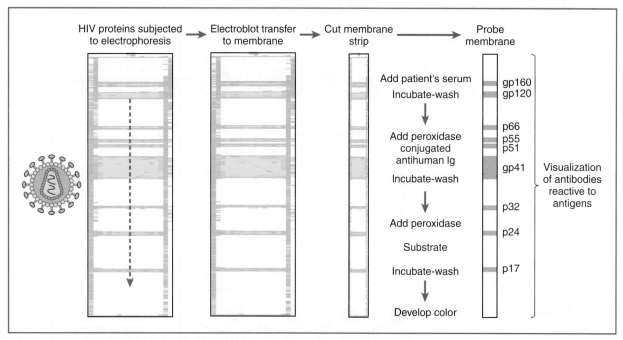

Figure 9-4. Western blot (immunoblotting). Reactivity to antigens can be assessed by Western blot analysis. An example of detection using serum antibodies is pictured. Human immunodeficiency virus (*HIV*) antigens separated by gel electrophoresis are transferred to a solid matrix (nitrocellulose), then probed with patient serum. Reactive antibodies against HIV antigens are visualized, thus determining patient exposure to the viral agent. *Ig*, immunoglobulin.

antigen-specific antibodies. Enzyme-linked immunosorbent assays (ELISAs), also known as *enzyme immunoassays* (EIAs) or *solid-phase immunoassays,* are designed to detect antigens or antibodies by producing an enzyme-triggered color change. ELISAs are referred to as solid-phase assays because they require the immobilization of antigens or antibodies on solid surfaces. The noncompetitive ELISA uses a specific antigen attached to a solid-phase surface (plastic bead or microtiter well) (Fig. 9-5). The test specimen is added; incubation allows antibodies in the test specimen to bind to the antigen. A secondary enzyme-labeled antibody is used to detect the test antibody. A chromogenic substrate is added; the enzyme converts the substrate to a detectable colored compound. The amount of color that develops is proportional to the amount of antibody in the test specimen.

A similar assay uses an antigen-specific antibody bound to the well. The test specimen is added, and any specific antigen is captured by the immobilized antibody. A second enzyme-linked antibody, which binds a different antigenic determinant on the antigen, is then added to detect captured antigen. In this format, the antigen is "sandwiched" between antibodies. Similarly to the noncompetitive ELISA, the amount of color produced in the presence of enzyme-labeled reporter is proportional to the antigen in the test specimen. Newer methodologies allow quantitative analytical detection of multiple antigens using similar experimental formats.

Immunohistochemical Methods

Antibody-based assays can be used to determine whether a particular antigen is found on or in the cells of a particular tissue. The method involves covalent attachment of reporter enzymes or fluorescent tags to specific antibodies to detect antigens. The enzyme horseradish peroxidase is frequently used as a coupled reagent, allowing deposition of diaminobenzidine for visualization of target cells (Fig. 9-6). Similarly, fluorochrome-tagged antibodies (fluorescein-, phycoerythrin-, rhodamine-, or allophycocyanin-conjugated antibodies) can be used to visualize antigen location after excitation with ultraviolet light.

Alternatively, a two-step method may be used in which unlabeled antibody specific for the antigen in question is reacted first with the tissue, followed by a tagged secondary antibody that recognizes the first specific antibody. Under these circumstances, the initial signal can be amplified, since multiple signals are generated by secondary antibodies that bind the primary antibody. The use of these antibodies with fluorescent tags, in concert with advances in microscopy, has transformed disease research. For example, immunofluorescence is widely used in clinical laboratories for screening patients' sera for anti-DNA antibodies in suspected cases of systemic lupus erythematosus.

Fluorescence-Activated Cell Sorting

Fluorescence-activated cell sorting analysis is used to identify, and sometimes purify, one cell subset from a mixture of cells (Fig. 9-7). Antibodies directed against surface molecules can be used to identify and quantitatively analyze specific cell phenotypes within a population of cells. Antibodies may be directly tagged with fluorescent molecules or indirectly identified with secondary tagging (indirect immunofluorescence). Labeled cells are passed single file through a laser beam by continuous flow in a fine suspension stream. Each cell scatters

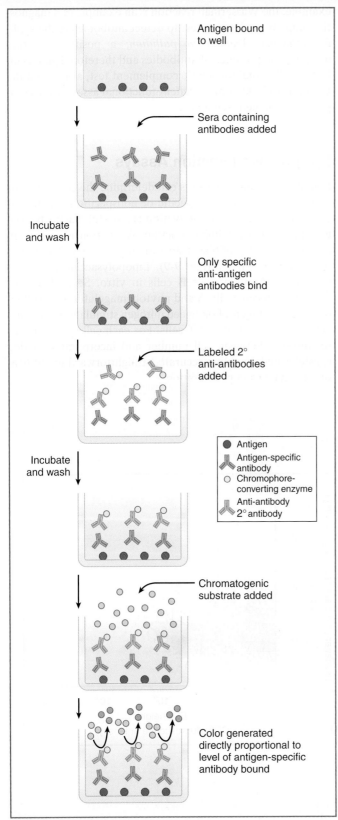

Figure 9-5. Enzyme-linked immunosorbent assay. Serum added to wells coated with antigen is probed with antibodies. Enzyme-conjugated secondary antibodies detect the primary complex. Catalytic conversion of chromogenic substrate allows quantitation against known standards.

some of the laser light. Detection of scattered light in a parallel manner (forward scatter) is indicative of cell size, whereas detection of scattered light at a perpendicular angle (side scatter) is indicative of cellular granularity. These physical parameters alone can distinctly identify granular polymorphonuclear cells, mononuclear cells, and RBCs. Antibody-labeled cells emit fluorescent light when excited by the laser. Single cells are detected at an extremely rapid rate of data acquisition. This technology is critical for defining populations of cells according to expression of surface molecules, or by their Cluster of Differentiation (CD) antigen designation. Newer methods allow detection of intracellular cytokines to match cytokine production within individual cell phenotypes.

ASSESSMENT OF IMMUNE FUNCTION

General Inflammation Assessment

Two general tests are effective at measuring inflammation due to disorder, infection, or tissue damage. The C-reactive protein test measures acute-phase reactant released by the liver. Elevated values are indicative of impending infection or inflammation, whether due to onset of sepsis or increased pathology due to disease activity. This is an excellent and simple way to monitor inflammation, especially before other clinical indicators. Likewise, another simple test, the erythrocyte sedimentation rate, is useful for indication of presence of inflammation. This test measures the rate at which RBCs precipitate in a specified period (usually 1 hour). It is a common hematology parameter that indicates the presence of high factors in blood that hide the overall negative charge present on erythrocytes (zeta potential). When an inflammatory process is present, the high protein (fibrinogen, immunoglobulins, proinflammatory cytokines) in the blood causes RBCs to stick to each other, forming a "rouleaux" stack that increases settlement speed.

Complement Fixation Test

The complement fixation test is a blood test in which a sample of serum is exposed to a particular antigen and complement to determine whether antibodies to that particular antigen are present (Fig. 9-8). The nature of complement is to react in combination with antigen-antibody complexes. The relative lack of antigen specificity allows complement to react with almost any antigen-antibody complex. In the test procedure, complement remains free unless "fixed" by the particular antigen and antibody system in question. The indicator used in many complement fixation assays is sheep RBCs. In a positive or reactive test, the complement is bound to an antigen-antibody complex and is not free to interact with target RBCs. The RBCs remain unlysed and settle to the bottom of a concave-shaped well to form a button. In a negative or nonreactive test, the complement remains free to interact with the blood cells, causing them to lyse. The complement test is a powerful tool to identify antibodies reacting with antigens to confirm exposure to a specific microorganism. For

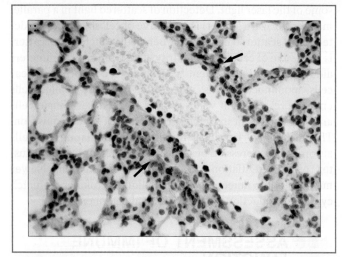

Figure 9-6. Immunohistochemical method. The use of antibodies to identify cell structure and phenotype within tissue is a powerful tool. Pictured are lymphocytes cuffing pulmonary vascular regions. Arrows indicate localized CD3+ lymphocytes in tissue. Enzyme (horseradish peroxidase)-conjugated antibodies directed against surface CD3+ were incubated with the tissue section. Conversion and subsequent deposition of chromogenic substrate (diaminobenzidine) enable identification and visualization of T lymphocytes.

example, the Wasserman reaction is an example of a diagnostic complement fixation test to detect antibodies to the syphilis organism *Treponema pallidum*; a positive reaction indicates the presence of antibodies and therefore syphilis infection. The total hemolytic complement test, also called the CH50 or CH100, is an excellent functional assay of the complete complement sequence.

Lymphocyte Function Assays

Lymphocyte function assays can determine the state of B- or T-cell responsiveness to specific or nonspecific antigens. Investigation of mitogenic activation is a useful diagnostic tool to establish basic cellular function. Activation is accurately measured by way of blast transformation assay as a measure of in vitro reactivity (Fig. 9-9). Lipopolysaccharides cause polyclonal stimulation of B cells in vitro. Several lectins, including concanavalin A and phytohemagglutinin, are effective T-cell mitogens. Pokeweed mitogen stimulates polyclonal activation of both B and T cells. For each cell type, a linear relationship between cell number and incorporated analog is established, enabling accurate, straightforward quantification of changes in proliferation.

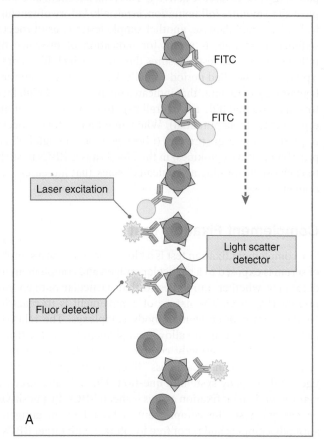
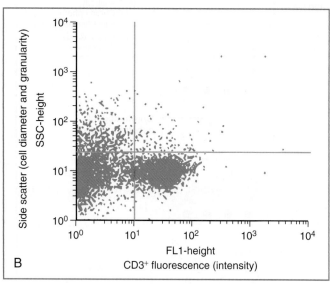

Figure 9-7. Fluorescence-activated cell sorting. Flow cytometric analysis allows direct identification of cell phenotypes using fluorescently tagged antibody molecules. Modern cytometers can measure multiple parameters simultaneously, including low-angle forward scatter (proportional to cell diameter), orthogonal scatter intensity (measuring cellular granularity), as well as fluorescence intensity Multiple cellular phenotypes may be detected using compounds that fluorescence at unique wavelenghts when excited. (**A**). Pictured is a scattergram of splenocytes screened for reactivity with an FITC fluorescein isothiocyanate labeled anti-CD3 antibody to allow quantitation of T lymphocytes within the population (**B**). A shift to the right on the x-axis indicates the presence of positively stained cells with increasing CD3 surface molecule present. *SSC*, side scatter.

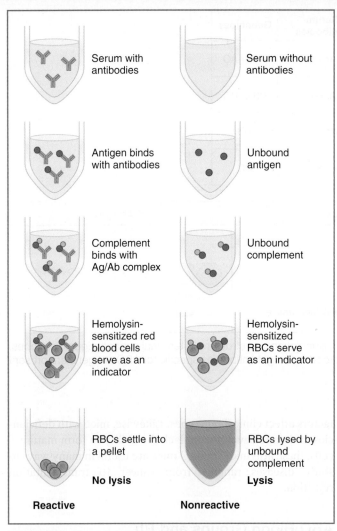

Figure 9-8. Complement fixation assay. In the complement fixation assay, complement components bind to antibody-antigen complexes (*Ag/Ab*), thereby making complement unavailable for hemolysis of indicator red blood cells (*RBCs*). In the absence of specific antibody-antigen interactions, complement assembly results in cell lysis.

Additional lymphocyte function assays include the mixed leukocyte reaction, in which inactivated recipient cells are mixed with donor lymphocytes to indicate whether CD4+ cells of a recipient individual react to class II major histocompatability complex of the donor. These assays are extremely powerful tools to assist in determination of transplant compatibility. Cytotoxicity assays measure the ability of cytotoxic T cells or natural killer cells to kill labeled target cells that express a specific antigen for which the cytotoxic T cells may be sensitive.

Microarrays to Assess Gene Expression

Levels of expression of thousands of genes can now be measured simultaneously using a technology called **gene chips** or **microarrays.** Briefly, thousands of short complementary DNA (cDNA) samples representing genes from all parts of the

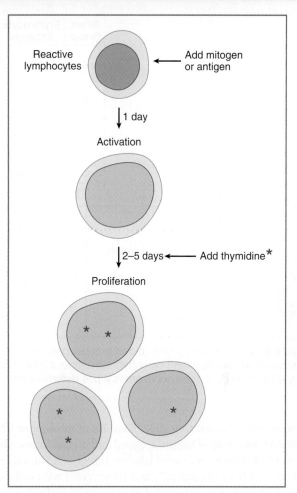

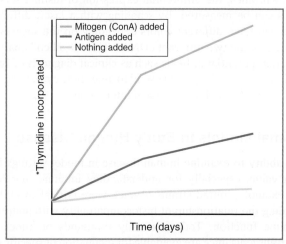

Figure 9-9. Blast transformation assays are useful to measure cell reactivity of lymphocytes. Peripheral blood lymphocytes are incubated in the presence of mitogen or specific antigen. If reactive, lymphocytes begin to proliferate. Historically, radioactive nucleotides (^{3}H-thymidine) were added; the amount of radioactivity incorporated into DNA was determined as a quantitative measure of proliferation. The reduction of tetrazolium salts is now recognized as a safer and accurate alternative to radiometric testing. The yellow tetrazolium salt is reduced in metabolically active cells to form insoluble purple formazan crystals, which are solubilized by the addition of a detergent. The color can then be quantified by spectrophotometric means. *ConA*, concanavalin A.

Blood Group	Erythrocyte Antigens	Serum Antibodies	Genotypes
A	A	Anti-B	AA or AO
B	B	Anti-A	BB or BO
AB	A and B	Neither	AB
O	Neither	Anti-A and anti-B	OO

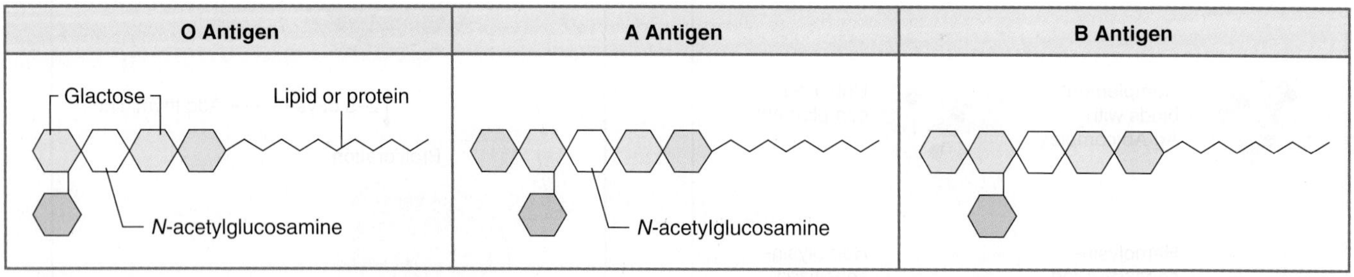

Figure 9-10. ABO phenotypes (blood groups) present in the human population and structure of terminal sugar antigenic epitopes. The blood groups represent carbohydrate antigens present on red blood cells. IgM antibodies specific for ABO antigens are present only if the individual does not express those determinants.

genome are attached to a slide. Samples of messenger RNA (mRNA) from cells in culture are used and reverse-transcribed into cDNA, and by means of labeling this cDNA from different sources (i.e., normal cells and tumor cells) with different fluorochromes, the differential expression of distinct sets of genes can be measured. By scanning with a laser, different spots can have different colors depending on the success of binding by the two different cDNAs. This "in silico" analysis has great potential in fields such as clinical diagnosis of lymphoid tumors, and has been useful to determine cytokine gene expression during host-pathogen interactions.

Animal Models to Study Human Disease

The ability to examine human disease in model settings has high value, especially for understanding immune function. For example, inbred strains of mice were critical to understanding the relationship of histocompatibility with adaptive immune function. Today, literally thousands of knockout and transgenic mice are available for evaluation of specific cytokines, receptors, and surface interactions, to understand the control and deregulation of response to antigens and pathogens. Cells of specific phenotype can be isolated, and used for adoptive transfer to genetically matched recipients to understand function during disease. Two specific models should be mentioned. Severe combined immunodeficiency disease arises when T and B lymphocytes fail to develop to allow adaptive function. For example, individuals with defects in the recombinase-activating genes (RAG) *RAG-1* or *RAG-2* can lead to a wide variety of immunodeficiencies with clinical complications. Severe combined immunodeficiency disease mice are critical as tools to decipher how developmental factors affect clinical outcomes. Likewise, mice with deficiencies in thymic development lack the ability to form mature T cells; these athymic ("nude") mice are useful in many ways to study diseases ranging from cancer to transplantation rejection.

ABO Blood Groups and Rh Incompatibility

The ABO blood groups were first identified in 1901. They represent important antigens to be accounted for to assure safe blood transfusions (Fig. 9-10). The ABO antigens represent carbohydrate moieties present on erythrocytes. Individuals naturally develop antibodies (called *isoantibodies*), usually of the IgM isotype, specific for ABO antigens that they do not express. If the individual receives a transfusion of blood that contains incompatible ABO antigens, isoantibodies will cause agglutination of the donor cells. This process is referred to as *isohemagglutination*; the antigens are sometimes called **isohemagglutinins**.

Rh antigens, also called **Rhesus antigens,** are transmembrane proteins expressed at the surface of erythrocytes. They appear to be used for the transport of CO_2 and/or ammonia across the plasma membrane. RBCs that are Rh positive express the one designated D (RhD antigen). About 15% of the population have no RhD antigens and thus are "Rh negative." Of great clinical importance is the complication of RhD incompatibility between mother and fetus. An Rh-negative mother who carries an Rh-positive fetus runs the risk of producing immune antibodies of the IgG isotype to the Rh antigens on the fetal RBC. The exposure during the primary pregnancy is minimized. However, the mother may generate Rh antibodies after giving birth if she comes into contact with fetal blood cells during placental rupture. Some fetal RBCs enter the mother's bloodstream, thereby allowing production of maternal derived anti-Rh antibodies. In subsequent pregnancies, the next Rh-positive fetus will be at risk, since the mother will retain a low level of circulating antibodies against the Rh antigen. Destruction of fetal erythrocytes will ensue by passive immune transfer of maternal antibodies to the fetus, resulting in erythroblastosis fetalis (hemolytic disease of the newborn). It is of great clinical importance to identify the Rh-mismatched mother and fetus; typically, an indirect Coombs test is performed to identify isohemagglutination. If it is positive, the mother is clinically treated with anti-Rh antibodies (Rh immune globulin or **RhoGAM**), which react with fetal RBCs. Ensuing antibody-antigen complexes are removed prior to maternal recognition of foreign Rh antigen.

KEY POINTS ABOUT IMMUNOASSAYS

- Immune experimental systems allow diagnostic assessment of infection and pathogenic responses.
- Interactions of antibodies with antigens form the basis of many qualitative and quantitative diagnostic assays.
- The affinity of the antibody-antigen reaction can be defined through physical laws of mass action.

KEY POINTS

- Interactions between antibodies and antigens are dictated by equilibrium constants within the host environment. These physical interactions form the basis of quantitative and qualitative assays, and work via the laws of mass action.
- Precipitation and agglutination assays, exemplified by the Coombs reaction, allow determination of reactivities in both solid and liquid mediums.
- Monoclonal and polyclonal antibodies are well suited as diagnostic reagents and allow precipitin and capture reactions to be detected in various formats.
- In addition to antibody-based reactions, clinical tests employ detection methodologies for complement component activation, for gene expression, and for lymphocytic reactivity. All these tools are critical to understand the interactions of immune function to detect and successfully defend against pathogenic infection.
- ABO phenotypes determined by antibody reactivity to carbohydrate antigenic epitopes on RBCs dictate the parameters for blood transfusions. Likewise, antibody reactivity determines the extent of Rh incompatibility between mother and fetus, and must be addressed during pregnancy to eliminate the chance of erythroblastosis fetalis (hemolytic disease of the newborn).

Self-assessment questions can be accessed at www.StudentConsult.com.

Infection and Immunity 10

CONTENTS
MAJOR IMMUNE DEFENSE MECHANISMS AGAINST PATHOGENS
Bacterial Infections
Mycobacterial Infections
Viral Infections
Parasitic Infections (Helminths)
Fungal Infections
IMMUNE DEVIATION: DOMINANCE OF ONE RESPONSE OVER ANOTHER
EVASION OF IMMUNE RESPONSE

●●● MAJOR IMMUNE DEFENSE MECHANISMS AGAINST PATHOGENS

The course of response against typical acute infections can be subdivided into distinct stages. Initially, the level of infectious agent is low, beginning with breach of a mechanical barrier (e.g., skin, mucosal surface). Once inside the host, the pathogen encounters a microenvironment for suitable replication. The agent replicates, releasing antigens that trigger innate immune function, generally characterized as nonspecific. Preformed effector molecules recognize microorganisms within the first 4 hours of infection and assist in limiting expansion of the organism. Complement components and released chemokines attract professional phagocytes (macrophages and polymorphonuclear [PMN] cells) and natural killer (NK) cells to the site of infection, to assist in activation of these cells. After 4 or 5 days, antigen-specific lymphocytes (B and T cells) undergo clonal expansion, enabling directed control and eventual clearance of the infectious agent. As the agent is cleared, the host is left with residual effector cells and antibodies as well as immunologic memory to provide lasting protection against reinfection.

A wide variety of pathogenic microorganisms exist. They may be globally classified into groups: bacteria, mycobacteria, viruses, protozoa, parasitic worms, and fungi. The major immune defense mechanisms are summarized in Table 10-1.

The host defense is based upon availability of resources to combat a localized pathogen (Fig. 10-1). Virtually all pathogens have an extracellular phase during which they are vulnerable to antibody-mediated effector mechanisms. An extracellular agent may reside on epithelial cell surfaces, where antibodies (immunoglobulin A) and nonspecific inflammatory cells may be sufficient for combating infection. If the agent resides within interstitial spaces, in blood, or in lymph, then protection may also include complement components and macrophage phagocytosis and neutralization responses. Intracellular agents require a different response to be effective. For cytoplasmic agents, T lymphocytes and NK cells, as well as T-cell–dependent macrophage activation, are usually necessary to kill the organism.

Pathogens can damage host tissue by direct and indirect mechanisms. Organisms may directly damage tissue by release of exotoxins that act on the surface of host cells or via released endotoxins that trigger local production of damaging cytokines. Pathogens may also directly destroy the cells they infect. Indirect damage can also occur through actions of the adaptive immune response. Pathologic damage may occur from excess deposition of antibody-antigen complexes or through bystander killing effects during overactive specific responses to infected host target cells.

Bacterial Infections

Bacterial infections begin with a breach of a mechanical barrier. Release of bacterial factors upon replication initiates a cascade of events (Fig. 10-2). Initially, infection may be resisted by antibody-mediated immune mechanisms, including neutralization of bacterial toxins. With the help of complement factors, direct cytotoxic lysis of microorganisms can occur. Release of C3a and C5a in the complement cascade causes vasodilatation and vasopermeability, resulting in an influx of professional phagocytes and acute PMN infiltrates. Opsonization of bacteria leads to increased phagocytosis and acute anaphylactic vascular events that permit exudation of inflammatory cells and fluids. Phagocytosis may also occur via specific surface receptors for ligands such as mannose or sialic acids (pattern recognition receptors). During the chronic stage of the infection, cell-mediated immunity (CMI) is activated. T cells reacting with bacterial antigens may infiltrate the site of infection, become activated, and release lymphokines that further attract and activate macrophages. Likewise, NK cells enter the infected region and assist in macrophage activation. The activated macrophages phagocytose and degrade necrotic bacteria and tissue, preparing the lesion for healing. PMNs, especially neutrophils, are an excellent example of the first line of innate defense against bacterial

TABLE 10-1. Major Immune Defense Mechanisms Against Pathogens

TYPE OF INFECTION	MAJOR IMMUNE DEFENSE MECHANISMS
Bacterial	Antibody, immunocomplex, and cytotoxicity
Mycobacterial	DTH and granulomatous reactions
Viral	Antibody (neutralization), CTL, and T_{DTH}
Protozoal	DTH and antibody
Parasitic worms	Antibody (atopic, ADCC) and granulomatous reactions
Fungal	DTH and granulomatous reactions

ADCC, antibody-dependent cell-mediated cytotoxicity; CTL, cytotoxic T lymphocyte; DTH, delayed-type hypersensitivity.

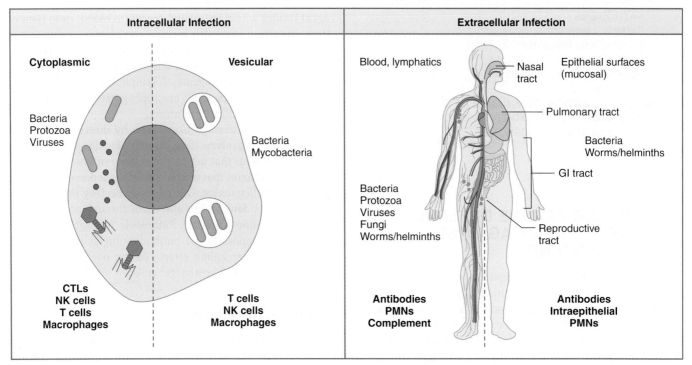

Figure 10-1. Functional immune response is dependent on organism location within the host. Effective immune responses are directed against intracellular organisms residing in cytoplasmic or vesicular space, or against extracellular organisms residing at mucosal surfaces or present in blood, lymph, or tissue. CTLs, cytotoxic T lymphocytes; NK, natural killer; PMNs, polymorphonuclear cells.

agents. Pus is composed of dead and dying PMNs and host cells, local fluids and exudates, and dead and dying bacteria.

The role of complement in response to bacterial infection must be stressed. Major biologic components of the complement system include activation of phagocytes, direct cytolysis of target cells, and opsonization of microorganisms and immunocomplexes for cells expressing complement receptors.

Important factors released by macrophages in response to bacterial antigens include cytokines that exert both local and systemic function. Locally, interleukin (IL)-1, tumor necrosis factor (TNF)-α, and IL-8 cause inflammation and activate vascular endothelial cells to increase permeability and allow more immune cells to enter infected area. TNF-α will also destroy local tissue to limit growth of bacteria. In addition, IL-6 can stimulate an increase in B-cell maturation and antibody production, and IL-12 will lead to activation of NK cells and priming of T cells toward a T helper cell (T_H) 1 response. Systemically, IL-1α, IL-1β, TNF-α, IL-6, and IL-8 all contribute to elevated body temperature (fever) and production of acute-phase proteins.

PHARMACOLOGY

Pharmacologic Control of Inflammation

Pharmacologic control of inflammation may be managed through the use of nonsteroidal antiinflammatory drugs (NSAIDs) or corticosteroid hormone analogs (such as prednisone). The NSAIDs as a group function as nonselective inhibitors of cyclooxygenase-1 (COX-1) and cyclooxygenase-2 (COX-2), which catalyzes the formation of prostaglandins and thromboxane from arachidonic acid.

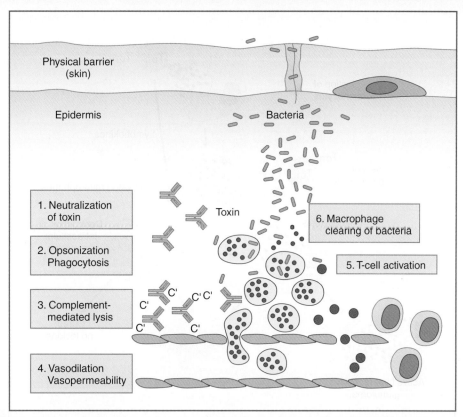

Figure 10-2. Bacterial infections. Immune defenses against bacterial agents include antibodies for neutralization of toxins, opsonization of organisms for targeted destruction, and activation of complement for direct lysis. Vasodilation of blood vessels allows entry of polymorphonuclear cells, macrophages, and T cells to sites of infection to assist in control of infection.

Mycobacterial Infections

Mycobacterial infections such as tuberculosis and leprosy are extremely complex (Fig. 10-3). Mycobacteria have evolved to inhibit normal macrophage killing mechanisms (e.g., phagosome-lysosome fusion) and survive within the "disarmed" professional phagocyte. T cell (delayed-type hypersensitive) initiated by T_H cells exhibiting delayed-type hypersensitivity (T_{DTH}) are the main mechanism involved in an active immune response, including granulomatous hypersensitivity, but only after infection has become established. Both helper and cytotoxic T cells are responsible for control and containment of active infection; NK T cells (NKTs) also play a role in initial response. These cells recognize mycobacterial antigens and glycolipids presented on the surface of infected cells, leading to release of cytokines and chemokines that recruit additional immune effectors. A local environment is established to contain infection, resulting in granuloma pathology. Healing of the infected center may occur, with limited necrosis of the infected tissue. If the infection persists, an active caseous granuloma may form with a necrotic nidus composed of infected and active macrophages. The granuloma is usually circumscribed by responding T cells, with host-mediated destructive response occurring inside the area of organism containment. The presence of giant cells (activated syncytial multinucleated epithelioid cells) is characteristic of late-stage response.

At one time it was thought that the tissue lesions of tuberculosis required the effect of delayed-type hypersensitivity. The term **hypersensitivity** was coined because animals with cellular immune reactivity to tubercle bacilli developed greater tissue lesions after reinoculation of bacilli than did animals injected for the first time. The granulomatous lesions seen in tuberculosis depend upon primary innate functions as well as acquired immune mechanisms; lesions are not the cause of disease but an unfortunate effect of protective mechanisms. In the lung, extensive damage with accompanying caseous granulomatous pathology can ultimately result in respiratory failure. The granulomatous immune response produces the lesion, but the mycobacterium causes the disease.

Viral Infections

Immune responses to viral agents are dependent upon location of the virus within the host (Fig. 10-4). Antibodies play a critical role during the extracellular life cycle of the virus. Antibodies can bind to virus-forming complexes that can inactivate virions and allow them to be cleared effectively by professional phagocytes. Humoral responses can prevent the entry of virus particles into cells by interfering with the ability of the virus to attach to a host cell, and secretory immunoglobulin A can prevent the establishment of viral infections of mucous membranes. Once viral infection is established within cells, it is no longer susceptible to the effects of antibody. Upon entry to cells, immune resistance to viral infections is primarily T-cell–mediated. To be effective in attacking intracellular organisms, an immune

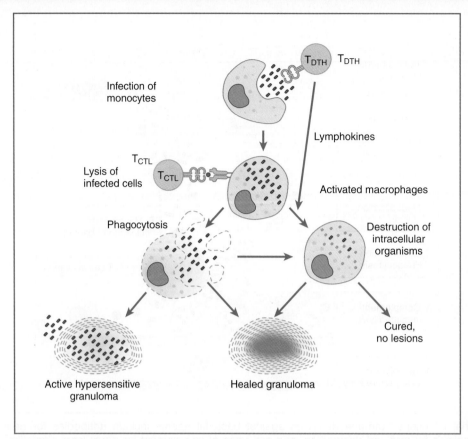

Figure 10-3. Mycobacterial infections. Immunity against mycobacteria is initiated by phagocytic macrophages, the preferred host for the infectious agent. The overall outcome and associated pathologies are dependent upon level of activation by delayed-type hypersensitivity (T_{DTH}) responses. *CTL*, cytotoxic T lymphocyte.

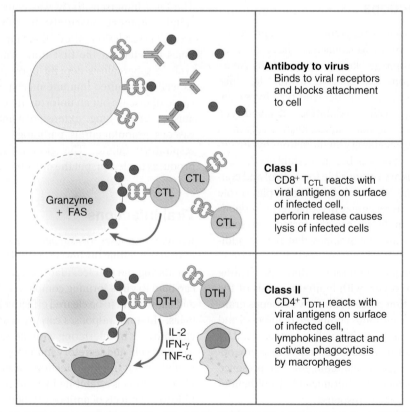

Figure 10-4. Viral infections. Immunity against viral infections is threefold, with contributions by antibodies, cytotoxic T cells (*CTL*), and T helper cells. *DTH*, delayed type hypersensitivity; *IL*, interleukin; *IFN*, interferon; *TNF*, tumor necrosis factor.

mechanism must have the capacity to react with cells in solid tissue. This is a property of cell-mediated reactions—particularly with cytotoxic T lymphocytes (CTLs)—but not of antibody-mediated reactions.

Most nucleated cells have an inherent, but limited, mechanism to downregulate viral replication through self-production of interferon (IFN)-α and IFN-β. However, these nonspecific interferon responses are not sufficient to eliminate the virus. NK cells are an early component of the host response to viral infection. NK cells nonspecifically recognize and kill virally infected targets. NK cells provide a link between innate and adaptive immunity, since they produce multiple immunomodulatory cytokines (e.g., IFN-γ, TNF-α, transforming growth factor-β [TGF-β], and IL-10). In addition, NK cells release IFN-γ (physically different from the other interferons) and IL-12, molecules that both activate macrophages and help prime T cells for an effective antiviral T_H1 response.

Virus-infected cells will, at some stage of the infection, express viral antigens on the cell surface in combination with class I molecules. Specific sensitized CD8+ CTL cells recognize presented viral antigens and destroy virus-infected cells (and therefore limit viral replication) through release of factors that include granzymes, perforins, and interferons. Lethal signals may be delivered through Fas/Fas-ligand–mediated mechanisms. Adverse effects occur if the cell expressing the viral antigens is important functionally, as is the case for certain viral infections of the central nervous system. If the virus-infected target is a macrophage, lymphocyte T_{DTH} cells can activate the macrophages to kill their intracellular viruses; lymphokine-activated macrophages produce a variety of enzymes and cytokines that can inactivate viruses. The critical nature of T-cell–mediated response to viral infections is evident in patients whose CMI is defective.

PATHOLOGY

Granulomatous Response to *Mycobacterium tuberculosis*

Cell-mediated response to mycobacterial antigens results in disruption of normal pulmonary architecture and development of a granulomatous response. Late-stage disease is characterized by large granulomas with central caseation, a process of necrosis that includes elements of liquefactive and coagulative necrosis. Cavities near the lung apex can allow entry of liquefied necrotic material into eroded airways, leading to spread of infectious agents via aerosolization.

Parasitic Infections (Helminths)

Host responses to parasitic worm infections are generally more complex because the pathogen is larger and not able to be engulfed by phagocytes (Fig. 10-5). Helminths typically undergo life cycle changes as they adapt for life in the host. Worms are located in the intestinal tract or tissues. Tapeworms, which exist in only the intestinal lumen, promote no protective immunologic response. On the other hand, worms with larval forms that invade tissue typically stimulate an immune response. The tissue reaction to *Ascaris* and *Trichinella* consists of an intense infiltrate of PMN leukocytes, with a predominance of eosinophils. Therefore, a variety of antigens that are life cycle stage dependent are displayed in changing tissue environments. Numerous cells play a role, depending on the location of the organism. Antigens on the surface of organisms or released into the local environment may stimulate T cells and macrophages to interact with B cells to secrete specific antibodies. IL-5, a T cell–derived factor, is instrumental in stimulation of eosinophils; the eosinophils act by associating with specific antibody to kill worms by antibody-dependent cell-mediated cytotoxicity (ADCC) or by releasing enzymes from granules to exert controlling effects on mast cells.

Antigen reacting with IgE antibody bound to intestinal mast cells stimulates release of inflammatory mediators, such as histamine, proteases, leukotrienes, prostaglandins, and serotonin. These agents cause an increase in the vascular permeability of the mucosa, exposing worms to serum immune components, and stimulate increased mucus production and increased peristalsis. These activities are associated with expulsion of parasitic worms from the gastrointestinal tract through formation of a physical barrier to limit adherence and interactions with the mucosal surface.

Eosinophil granules contain basic proteins that are toxic to worms. Eosinophils may be directed to attack helminths by cytophilic antibodies that bridge the eosinophil through the Fc region and the helminth by specific Fab-binding ADCC. Anaphylactic antibodies (IgE) are frequently associated with helminth infections, and intradermal injection of worm extracts elicits wheal-and-flare reactions. Children infested with *Ascaris lumbricoides* have attacks of urticaria, asthma, and other anaphylactic or atopic reactions that are presumably associated with dissemination of *Ascaris* antigens.

PATHOLOGY

Urticaria

Urticaria, or hives, resulting from localized mast cell degranulation, are wheal-and-flare reactions on the skin surface, with accompanying angioedema in deeper layers of the dermis.

Fungal Infections

Relatively little is known about the immune response to fungal agents. Cellular immunity appears to be the most important immunologic factor in resistance to fungal infections, although humoral antibody certainly may play a role. T_H1 type responses are protective via release of IFN-γ. By contrast, T_H2 responses (IL-4 and IL-10) typically correlate with disease exacerbation and pathology. The importance of cellular reactions is indicated by the intense mononuclear infiltrate and granulomatous reactions that occur in tissues infected with fungi and by the fact that fungal infections are frequently associated with depressed immune reactivity of the delayed type (opportunistic infections). For example, the condition of chronic

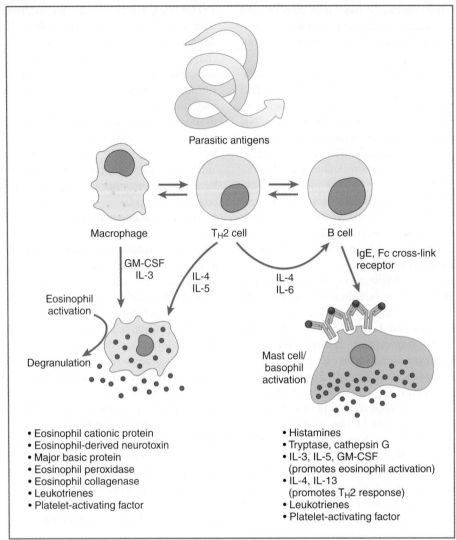

Figure 10-5. Response to parasitic worms. Immune activity against parasitic worms is directed by T helper 2 (T_H2) cells, driving activation of eosinophils, basophils, and mast cells to release inflammatory mediators to limit parasitic activity and kill invading organisms. Specific responses can be directed depending on use of either high-affinity Fcε receptors present on mast cells, or lower-affinity Fcε receptors present on eosinophils. *GM-CSF*, granulocyte macrophage colony-stimulating factor; *IL*, interleukin; *Ig*, immunoglobulin.

mucocutaneous candidiasis caused by persistent or recurrent infection by *Candida albicans* usually manifests only in patients with general depression of cellular immune reactions. As a general rule, fungi appear to be resistant to the effects of antibody, and CMI is needed for effective resistance.

KEY POINTS ABOUT MAJOR IMMUNE DEFENSE MECHANISMS AGAINST PATHOGENS

- The host defense is based upon availability of resources to combat a localized pathogen.
- Virtually all pathogens have an extracellular phase during which they are vulnerable to antibody-mediated effector mechanisms and complement components, macrophage phagocytosis, and neutralization responses.
- Intracellular agents usually require T lymphocytes (helper and cytotoxic) and NK cells, as well as T-cell–dependent macrophage activation, to kill organisms.
- Pathogens can damage host tissue by direct and indirect mechanisms.
- The main immune mechanisms against pathogens are: **bacterial,** antibody (immunocomplex and cytotoxicity); **mycobacterial,** DTH and granulomatous reactions; **viral,** antibody (neutralization), CTL, and T helper; **protozoal,** DTH and antibody; **parasitic worms,** antibody (atopic, ADCC) and granulomatous reactions; **fungal,** DTH and granulomatous reactions.

●●● IMMUNE DEVIATION: DOMINANCE OF ONE RESPONSE OVER ANOTHER

Immune deviation, or split tolerance, is defined as the dominance of one immune response mechanism over another for a specific antigen and has been implicated in the tendency for certain individuals to develop IgE (allergic) antibodies rather than IgG antibodies. In addition, for reasons that may be genetically determined, some individuals tend to make strong cellular immune responses but weak antibody response to certain antigens, whereas other individuals will have the opposite responses.

The biologic function of granulomatous reactivity is exemplified by the spectrum of pathology presented during infection with *Mycobacterium leprae*, the causative agent for leprosy. The clinical manifestations of leprosy are determined by the immune response of the patient. Leprosy is classified into three major overlapping groups: tuberculoid, borderline, and lepromatous. Tuberculoid leprosy is associated with delayed hypersensitivity and the formation of prominent, well-formed granulomatous lesions, many lymphocytes, and few if any organisms. Delayed hypersensitivity skin tests are intact, and there is predominant hyperplasia of the diffuse cortex (T-cell zone) of the lymph nodes. The low resistance characteristic of lepromatous leprosy is associated with the accumulation of "foamy" macrophages filled with viable organisms and the presence of high levels of humoral antibodies. Delayed hypersensitivity skin tests are depressed, and there is marked follicular hyperplasia in the lymph nodes with little or no diffuse cortex. The levels of antibodies are high, and vascular lesions due to immunocomplexes are seen (erythema nodosum leprosum). Borderline leprosy has intermediate findings.

This example of the forms of leprosy illustrates both the role of cellular immunity (delayed hypersensitivity) in controlling the infection and the lack of protective response provided by humoral antibodies. The cytokine patterns in the two polar forms of the disease are different. Typically T_H2 cytokines (IL-4, IL-5, and IL-10) dominate in the lepromatous form, whereas cytokines produced by T_H1 cells (IFN-γ, TNF, and IL-2) predominate in tuberculoid leprosy. IFN-γ would be expected to activate macrophages to kill intracellular pathogens and control organism expansion; high IL-4 may explain hypergammaglobulinemia in lepromatous patients. A diagram illustrating the relationship of the degree of cellular and humoral immune response to the stages of leprosy is shown (Fig. 10-6).

KEY POINTS ABOUT IMMUNE DEVIATION

- The response to initial infection is divided into phases.
- The first phase is an early innate and nonspecific response, in which preformed effector cells and molecules recognize microorganisms.
- The next phase is again primarily a nonspecific encounter with the organism, characterized by recruitment of professional phagocytes and NK cells to the site of infection.
- The final phase involves antigen-specific cell (B and T lymphocyte) effectors that undergo clonal expansion; these cells provide memory responses in case of reinfection.

EVASION OF IMMUNE RESPONSE

In the ongoing evolution of host-parasite relationships between humans and their infections, infectious organisms have developed "ingenious" ways to avoid immune defense mechanisms (Table 10-2). Organisms may locate in niches (privileged sites) not accessible to immune effector mechanisms (**protective niche**) or hide themselves by acquiring host molecules (**masking**). They may change their surface antigens (**antigenic modulation**), hide within cells, and produce factors that inhibit the immune response

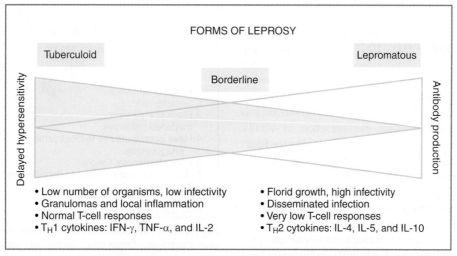

Figure 10-6. Immune deviation in response to *Mycobacterium leprae*. The *overlapping triangles* indicate the relative strength of delayed hypersensitivity and antibody production. The *solid triangle* indicates delayed hypersensitivity; the *open triangle*, antibody production. High levels of delayed-type hypersensitivity are associated with cure of tuberculoid leprosy; weak delayed-type hypersensitivity is associated with progressive disease; balanced DTH and antibody production with borderline leprosy and slowly progressive disease. T_H, T helper cell; *IFN*, interferon; *TNF*, tumor necrosis factor; *IL*, interleukin.

TABLE 10-2. Pathogen Evasion of Immune Response

MECHANISMS	EXAMPLES
Localization in protective niches	Latent syphilis, tapeworm (*Echinococcus*)
Intracellular location	Histoplasmosis, herpes virus, varicella, HIV
Antigenic modulation	Malaria, trypanosomiasis, relapsing fever
Preservation of receptor sites after reaction with antibody	Influenza virus
Immunosuppression	Malaria, measles, HIV, tuberculosis (anergy)
Inappropriate immune response (immune deviation)	Lepromatous leprosy, chronic mucocutaneous candidiasis

HIV, human immunodeficiency virus.

(**immunosuppression**) or fool the immune system into responding with an ineffective effector mechanism (**immune deviation**). The ultimate endpoint of coevolution of the human host and its infectious organisms results in an eventual mutual coexistence with most environmental organisms. No better evidence is the loss of this coexistence when the immune mechanisms do not function properly. Then, organisms that do not normally cause disease become virulent. The lesson of AIDS (infection with the human immunodeficiency virus) demonstrates that opportunistic infectious organisms will become dominant when introduced into a previously unexposed population. In a fully evolved, mature relationship, host and infectious agent initially coexist with limited detrimental affects. Thus, the ultimate evolution of the host-parasite relationship is not "cure" of an infection by complete elimination of the parasite, but at least mutual coexistence without deleterious effects of the parasite on the host. In fact, in many human infections, the infectious agent is never fully destroyed and the disease enters a latent state, only to be reactivated when immune surveillance wanes. For example, 70% of the population has been exposed to cytomegalovirus; transplant patients who are therapeutically immunosuppressed are at much greater risk for development of infection-related disease.

Bacteria have evolved to evade different aspects of phagocyte-mediated killing (Fig. 10-7). For example, they may (1) secrete toxins to inhibit chemotaxis, (2) contain outer capsules that block attachment, (3) block intracellular fusion with lysosomal compartments, and (4) escape from the phagosome to multiply in the cytoplasm. Viral entities also subvert immune responses, usually through the presence of virally encoded proteins. Some of these proteins block effector functions of antibody binding, block complement-mediated pathways, inhibit activation of infected cells, and can downregulate major histocompatability complex class I antigens to escape CTL killing. Herpes virus produces a factor that inhibits inflammatory responses by blocking effects of cytokines through receptor mimicking, and another that blocks proper antigen presentation and processing. Finally, Epstein-Barr virus encodes a cytokine homologue of IL-10 that leads to immunosuppression of the host by activating T_H2 rather than T_H1 responses.

KEY POINTS ABOUT EVASION OF IMMUNE RESPONSE

- Virtually all classes of infectious agents have devised ways to avoid host defenses.
- Mechanisms include inaccessibility in protective niches, antigenic modulation of surface molecules, and release of factors to either suppress the immune response or cause immune deviation and ineffective response to the agent.

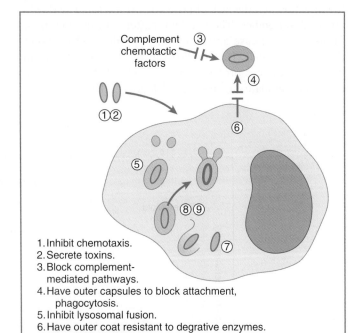

1. Inhibit chemotaxis.
2. Secrete toxins.
3. Block complement-mediated pathways.
4. Have outer capsules to block attachment, phagocytosis.
5. Inhibit lysosomal fusion.
6. Have outer coat resistant to degrative enzymes.
7. Escape from phagosome.
8. Turn off cytokine activation.
9. Activate cytokines inappropriately.

Figure 10-7. Mechanisms of infectious organisms to avoid immune defenses. Organisms evade immune responses through various mechanisms including location in protective niches, acquisition of host molecules, alteration of surface antigens, and producing factors to inhibit or redirect effective immune response.

KEY POINTS

- Immune defense against pathogenic organisms is tailored to meet the broad range of their extracellular and intracellular life-cycles within the host environment.
- Defense against bacterial agents primarily utilizes antibodies, antibodies and complement, and direct cytotoxic mechanisms to control infection.
- Defense against mycobacteria requires T-cell–mediated DTH responses that result in granuloma formation. Antifungal defenses also use similar mechanisms to control organisms.
- Defense against viral agents requires antibody neutralization upon initial infection, and cytotoxic mechanisms regulated by NK cells and CTLs when expanding within cellular compartments.
- Defense against protozoal agents incorporates DTH and antibody to limit growth.
- Defense against helminths and larger multicellular organisms utilizes atopic and ADCC-dependent reactions, as well as granulomatous responses, to sequester and destroy deposited eggs.
- Organisms have evolved multiple mechanisms to evade host responses, ranging from antigenic modulation of surface proteins to direct immunosuppressive action on specific cellular subsets.

Self-assessment questions can be accessed at www.StudentConsult.com.

SECTION II
Microbiology

SECTION 2

Microbiology

Basic Bacteriology 11

CONTENTS
BACTERIAL STRUCTURE, FUNCTION, AND CLASSIFICATION
GENERAL PROPERTIES OF PROKARYOTIC ORGANISMS
BACTERIAL PHYSIOLOGY
COMMENSAL ORGANISMS OF THE NORMAL BODY FLORA
BACTERIAL GENETICS
  Methods of Genetic Transfer Between Organisms
  Gene Expression and Regulation
BACTEREMIA, SYSTEMIC INFLAMMATORY RESPONSE SYNDROME, AND SEPSIS
BACTERIAL TOXINS: VIRULENCE FACTORS THAT TRIGGER PATHOLOGY
CLINICAL DIAGNOSIS
MAJOR ANTIMICROBIAL AGENTS AGAINST BACTERIA

At least 800 different species of bacteria inhabit the human host, representing a total population approaching 10^{15} organisms. Put into perspective, the number of bacteria is far greater than the number of cells in our bodies. Many organisms colonize various body tissues, representing specific flora that take advantage of space and nutrients in a commensal existence. However, organisms that forgo commensal or symbiotic relationships can produce disease and pathogenic response.

••• BACTERIAL STRUCTURE, FUNCTION, AND CLASSIFICATION

Historically, organisms were classified according to physical parameters, such as microscopic morphology (size and shape), staining characteristics, and ability to multiply on various energy sources (Fig. 11-1). Identification of specific biomarkers (biotyping) allowed classification for epidemiologic purposes, identifying organisms according to metabolic activity due to presence or absence of enzymes or ability to grow on specific substrates. The advent of antibiotics allowed further classification according to drug susceptibility patterns, and antibody-based serotyping was used to determine specific antigenic surface molecules unique to groups of bacterial organisms.

Recent development of molecular biologic tools has led to genotypic classification with greater precision than that of past methodologies. Genetic characterization of organisms is based directly on nucleic acid sequence and DNA homology, on nucleotide content (ratios of guanine plus cytosine), on analysis of plasmid content, or on ribotyping (RNA complement of a cell).

••• GENERAL PROPERTIES OF PROKARYOTIC ORGANISMS

Prokaryotic organisms have distinct characteristics from eukaryotes. Prokaryotic cells do not have a nuclear membrane; instead, their haploid circular DNA is loosely organized as a fibrous mass in the cytoplasm. Bacteria do not have organelles, unique Golgi apparatus, or true endoplasmic reticulum; rather, transcription and translation are coupled events. Bacterial 70 S ribosomes, consisting of 30 S and 50 S subunits, are significantly different from eukaryotic 80 S ribosomes, thus allowing potential targets for antimicrobials.

The cell envelope surrounding a bacterium includes a cell membrane and a peptidoglycan layer. Two major classes of bacteria are distinguishable by staining patterns following exposure to primary stain, gram iodine, and alcohol decolorization. **Gram-positive organisms** maintain a purple color from the primary stain incorporated into the thick peptidoglycan layer that surrounds the organism (Fig. 11-2). **Gram-negative organisms** have a reduced peptidoglycan layer surrounded by an outer membrane. The peptidoglycan layer is a complex polymer composed of alternating *N*-acetylglucosamine and *N*-acetylmuramic acid with attached tetrapeptide side chains. The bonds linking the *N*-acetylglucosamine and *N*-acetylmuramic acid are especially sensitive to cleavage by lysozyme, commonly found in saliva, tears, and mucosal secretions (useful basic host defense mechanisms). Gram-positive cell membranes are further characterized by the presence of teichoic and teichuronic acids (water-soluble polymers) chemically bonded to the peptidoglycan layer. Gram-negative bacteria are further characterized by the presence of a periplasmic space between the cell membrane and the outer membrane. The outer membrane is composed of a phospholipid bilayer with embedded proteins that assist in energy conversion (such as cytochromes and enzymes involved in electron transport and oxidative phosphorylation). The cytoplasmic membrane also contains enzymes critical for cell wall biosynthesis, phospholipid synthesis and DNA replication, and proteins that assist in transport of needed molecules.

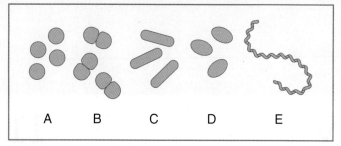

Figure 11-1. The diverse morphology of bacteria is related to physical characteristics of the outer cell membrane. Some of the diverse bacterial forms are cocci (*A*), diplococcic (*B*), bacilli (*C*), coccobacilli (*D*), and spirochetes (*E*).

Lipopolysaccharide (LPS) is contained within the outer membrane of gram-negative organisms and is composed of polysaccharide (O) side chains, core polysaccharides, and lipid A endotoxin. Lipid A contains fatty acids that are inserted into the bacterial outer membrane. The remaining extracellular portion of LPS is free to interact with host immune cells during infection, acting as a powerful immunostimulant via binding to the CD14 receptor on macrophages and endothelial cells and interactions via the TLR2 and TLR4 on cell surfaces, resulting in secretion of interleukins, chemokines, and inflammatory cytokines. Lipid core polysaccharides contain ketodeoxyoctonate as well as other sugars (e.g., ketodeoxyoctulonate and heptulose) and two glucosamine sugar derivatives. Lipoproteins link the thin peptidoglycan layer to an outer membrane.

Certain gram-positive and -negative organisms may also have a capsule, or glycocalyx layer, external to the cell wall, containing antigenic proteins. The capsule protects bacteria from phagocytosis by monocytes and can also play a role in adherence to host tissue. The glycocalyx is a loose network of polysaccharide fibers with adhesive properties containing embedded antigenic proteins. Alternatively, the outer wall may be composed of mycolic acids and other glycolipids,

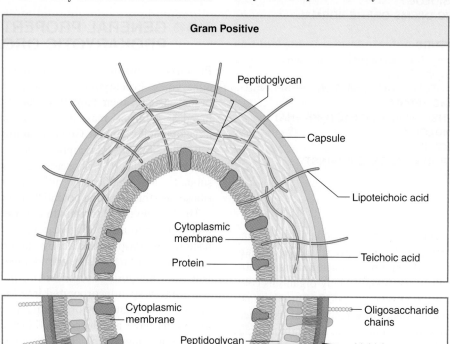

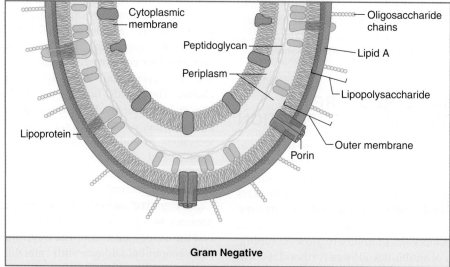

Figure 11-2. The gram-positive bacteria have a characteristic thick peptidoglycan layer surrounding an inner cytoplasmic membrane. The gram-negative bacteria have reduced peptidoglycan surrounded by periplasm, with an outer membrane comprised of embedded core lipopolysaccharide and lipid A endotoxin. Both gram-positive and gram-negative bacteria may also support an outer glycocalyx or capsule (not depicted). Upon treatment with gram iodine, gram-positive bacteria resist alcohol treatment and retain stain, whereas gram-negative organisms can be differentiated by loss of the primary stain and later addition of a safranin counterstain.

which provide extra protection during the process of host infection. Organisms can be further characterized by the presence of appendages, such as flagella, which assist in locomotion, or pili (fimbriae), which allow adhesion to host tissue; sex pili are involved in attachment of donor and recipient organisms during replication.

KEY POINTS ABOUT PROKARYOTIC ORGANISMS

- Bacteria are classified according to morphologic structure, metabolic activity, and environmental factors needed for survival.
- Gram-positive organisms are so named because of staining characteristics inherent in their thick peptidoglycan outer layer with teichoic and lipoteichoic acid present; gram-negative bacteria have a thin peptidoglycan component surrounding a periplasmic space, as well as an outer membrane with associated lipoproteins.
- Other bacterial species, such as mycobacteria, have unique glycolipids that give them a waxy appearance and unique biologic advantages during infection of the host.

BACTERIAL PHYSIOLOGY

All microorganisms of medical significance require energy obtained through exothermic reactions—chemosynthesis—and all require a source of carbon. Organisms capable of using CO_2 are considered autotrophs. Many pathogenic organisms are able to utilize complex organic compounds; however, almost all can survive on simple organic compounds such as glucose. The main scheme for producing energy is through glycolysis via the Embden-Meyerhof pathway (Fig. 11-3). Two other main sources for energy production are the tricarboxylic acid cycle and oxidative phosphorylation. Alternatively, the pentose-phosphate pathway may be used.

Facultative organisms can live under aerobic or anaerobic conditions. Obligate aerobes are restricted to the use of oxygen as the final electron acceptor. Anaerobes (growing in the absence of molecular O_2) use the process of fermentation, which may be defined as the energy-yielding anaerobic metabolic breakdown of a nutrient molecule, such as glucose, without net oxidation. Fermentation yields lactate, acetic acid, ethanol, or other simple products (e.g., formic acid).

Many bacteria are saprophytes, growing on decayed animal or vegetable matter. Saprophytes do not normally invade living tissue but rather grow in our environment. However, saprophytic organisms can become pathogenic in immunosuppressed individuals or in devitalized tissue, as seen with species of *Clostridium*.

KEY POINTS ABOUT BACTERIAL PHYSIOLOGY

- The three main schemes for producing energy in bacteria are glycolysis, the tricarboxylic acid cycle, and oxidative phosphorylation.
- Alternative schemes are used, such as fermentation, to assist organisms growing under anaerobic conditions.

COMMENSAL ORGANISMS OF THE NORMAL BODY FLORA

The body is host to a tremendous number of commensal organisms, existing in a symbiotic relationship in which a bacterial species derives benefit and the host is relatively unharmed (Fig. 11-4). The normal body flora consists of organisms that take advantage of the interface between the host and the environment, with the presence of defined species on exposed surfaces as well as throughout the respiratory, gastrointestinal, and reproductive tracts. Estimates are that a normal human body houses about 10^{12} bacteria on the skin, 10^{10} in the mouth, and 10^{14} in the gastrointestinal tract—numbers far in excess of eukaryotic cells in the entire body. The interaction between human host and residing normal body flora has evolved to the benefit of the host. For example, the normal flora prevents colonization of the body by competing pathogens and may even stimulate the production of cross-protective antibodies against invading organisms. In addition, the flora colonizing the gastrointestinal tract secretes excess vitamins (vitamins K and B_{12}) that benefit the host.

While the skin functions as a physical barrier to the outside world, it also serves as host for *Staphylococcus epidermidis* and *Corynebacterium diphtheriae*, both implicated in acne formation. The most important single mechanism for the purpose of keeping healthy is to frequently wash the hands to limit spread of both commensal and pathogenic bacteria. In the respiratory tract, nasal carriage is a primary niche for opportunistic *Staphylococcus aureus*, while the pharynx is commonly host to colonization by *Neisseria meningitidis*, *Haemophilus influenzae*, and streptococcal species. The upper respiratory tract commonly captures organisms in the mucosal layer, with subsequent clearance by cilia. The lower respiratory tract requires more aggressive methods for bacterial clearance, with heavy reliance upon macrophages and phagocytes to maintain a relative balance of bacterial cell numbers.

The oral cavity and gastrointestinal tract are host to a variety of organisms with physical properties allowing commensalism in these tissues. The mouth is host to more than 300 species of bacteria, while the stomach hosts fewer numbers of organisms that can survive the acidic environment (pH ≤ 2.0). The small bowel is relatively barren of organisms owing to fast-moving peristalsis, whereas the slower mobility of the large intestine allows residence of a high number of bile-resistant enteric pathogens (e.g., *Bacteroides* species), thus producing large numbers of aerobes and facultative anaerobes in the stool.

The reproductive organs are also host to a variety of organisms; *Escherichia coli* and group B streptococcus are commonly associated with vaginal epithelium, where they exist under conditions of high acidity. Indeed, *Lactobacillus acidophilus* colonizes the vaginal epithelium during childbearing years and helps establish the low pH that inhibits the growth of other pathogens. Expansion of organisms then occurs readily in postmenopausal women, who lose general acidity of this tissue.

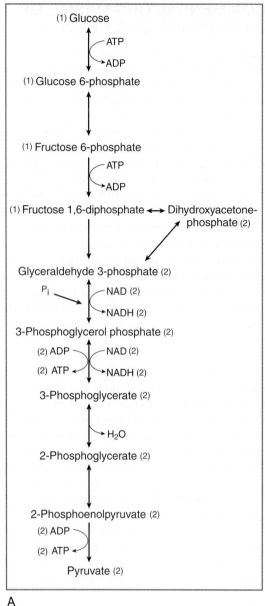
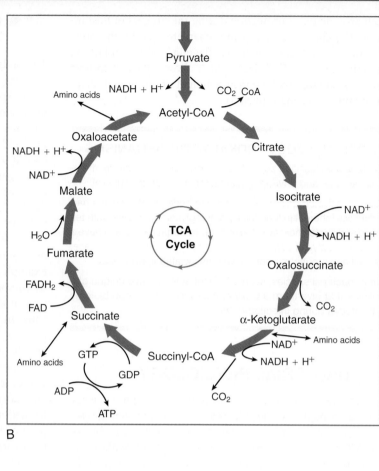

Figure 11-3. Two of the three main energy producing schemes used by bacteria for production of energy are the Embden-Meyerhof pathway (glycolysis) (**A**) and the use of pyruvate through the tricarboxylic acid (*TCA*) cycle to produce reduced nicotinamide adenine dinucleotide (*NADH*), reduced form of flavin adenine dinucleotide (*FADH$_2$*), and adenosine diphosphate (*ADP*) (**B**). A third method (not shown) utilizes oxidative phosphorylation through the electron transport chain. *ATP,* adenosine triphosphate; *CoA,* coenzyme A; *GTP,* guanosine triphosphate; *GDP,* guanosine diphosphate; *P$_i$,* inorganic phosphate.

BIOCHEMISTRY

Pentose-Phosphate Pathway

The pentose-phosphate pathway, also called the hexose monophosphate shunt, is an alternative mode of glucose oxidation that is coupled to the formation of reduced coenzyme reduced nicotinamide adenine dinucleotide phosphate, giving rise to phosphogluconate. The production of biosynthetic sugars is regulated by transketolases and transaldolases. In eukaryotic cells, the phosphogluconate path is the principal source of reducing power for biosynthetic reactions in most cells.

HISTOLOGY

Respiratory Cilia

Ciliated cells throughout conducting airways continue into the respiratory tract well past mucus-producing cells, in essence preventing mucus accumulation. Goblet cells are the unicellular mucous glands, aptly named since mucus accumulation in the apical portion gives the appearance of a goblet globe and the compressed basal portion of the cytoplasm gives the appearance of a goblet stem. Mucus stains poorly with hematoxylin and eosin. The abundant concentration of goblet cells decreases progressively in the bronchial passages, and they are completely absent from terminal bronchiole epithelium.

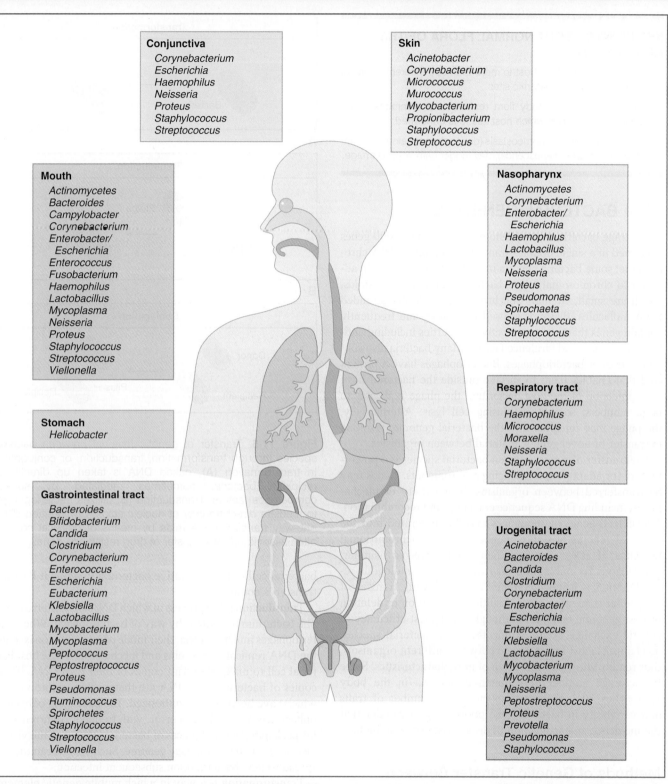

Figure 11-4. Tissue tropisms for commensal bacterial flora. The normal body flora represents organisms with tropism for specific anatomic sites. While internal tissues remain relatively free of bacterial species, tissue that is in contact with the environment, whether directly or indirectly, can be readily colonized. Some of the more common bacteria associated with specific anatomic sites of the human host are listed.

KEY POINTS ABOUT NORMAL FLORA OF THE HUMAN BODY

- The human body plays host to more than 200 different bacterial species at multiple anatomic sites.
- Much of the normal body flora resides in a commensal and mutual relationship in which host tissue is unharmed.
- However, alterations in homeostasis (due to stress, malnutrition, immune suppression, senescence) may trigger pathologic damage.

●●● BACTERIAL GENETICS

An average bacterium has a genome composed of 3000 genes contained in a single double-stranded, supercoiled DNA chromosome; some bacteria contain multiple chromosomes. In addition to chromosomal DNA, bacteria may harbor plasmids, which are small, circular, nonchromosomal, double-stranded DNA molecules. Plasmids are self-replicating and frequently contain genes that confer protective properties including antibiotic resistance and virulence factors. Many bacteria also contain viruses or bacteriophages. Bacteriophages have a protein coat that enables them to survive outside the bacterial host; upon infection of the host bacterium, the phage replicates to large numbers, sometimes causing cell lysis. Alternatively, the phage may integrate into the bacterial genome, resulting in transfer of novel genetic material between organisms.

The transfer of genes between bacterial species is a powerful tool for adaptation to changing environments. Genes may be transferred between organisms by a variety of mechanisms, including DNA sequence exchange and recombination. Transferred genes or sequences may be integrated into the bacterial genome or stably maintained as extrachromosomal elements. If DNA sequences being transferred are similar, homologous recombination may occur. Nonhomologous recombination is a more complex event allowing transfer of nonsimilar sequences, often resulting in mutation or deletion of host genomic material. Although the highest efficiency of genetic exchange occurs within the same bacterial species, mechanisms exist for exchange between different organisms, thus readily allowing acquisition of new characteristics. Since the average number of commensal bacteria in the body approaches 10^{14}, there are an enormous number of traits and variability in bacterial gene pools. It is no wonder that the incidence and acquisition of drug resistance is so high.

Methods of Genetic Transfer Between Organisms

The three main ways to transfer genes between bacterial organisms are conjugation, transduction, and transformation (Fig. 11-5).

Bacterial **conjugation** is the bacterial equivalent of sexual reproduction or mating. To perform conjugation, one bacterium has to carry a transferable plasmid (referred to as either an *F+* or an *R+ plasmid*), while the other must not. The transfer of plasmid DNA occurs from the F-positive

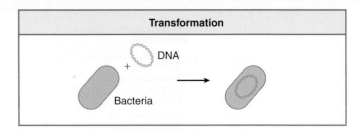
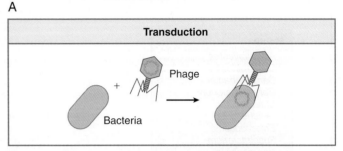
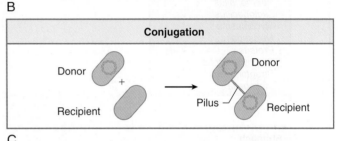

Figure 11-5. Transfer of genetic material between bacterial species through transformation, transduction, or conjugation. In transformation (**A**), naked DNA is taken up directly by recipient bacteria, sometimes mediated by surface competence factors. Transduction (**B**) utilizes bacteriophages to mediate direct transfer of nucleic acids. Conjugation (**C**) is one-way transfer of plasmids by means of physical contact, often associated with transfer of drug resistance genes.

bacterial cell to the F-negative bacterium (making it F+ once transfer is complete).

Transduction is the process in which DNA is transferred from one bacterium to another by way of bacteriophage. When bacteriophages infect bacteria, their mode of reproduction is to use the DNA replication proteins and mechanisms of the host bacterial cell to make abundant copies of their own DNA. These copies of bacteriophage DNA are then packaged into virions, which have been newly synthesized. The packaging of bacteriophage DNA is subject to error, with frequent occurrences of mispackaging of small pieces of bacterial DNA into the virions instead of the bacteriophage genome. Such virions can then be spread to new bacteria upon subsequent infection.

Transformation is a way in which mobile genetic elements move around to different positions within the genome of a single cell. Transposons are sequences of DNA, also called *jumping genes* or *transposable genetic elements*, that move directly from one position to another within the genome. During transformation, the insertion of sequences can both cause mutation and change the amount of DNA in the genome.

Bacteria multiply by binary fission. Figure 11-6 shows a model of bacterial growth, with growth rate directly tied to levels of nutrients in the local environment. The rate of bacterial growth is also dependent upon the specific organism; *E. coli* in nutrient broth will replicate in 20 minutes, whereas *Mycobacterium tuberculosis* has a doubling time of 28 to 34 hours. Bacterial DNA replication is a sequential three-phase process that uses a variety of proteins (Fig. 11-7). Initiation of replication begins at a unique genetic site, referred to as the origin of replication. Chain elongation occurs in a bidirectional mode. The addition of nucleotides occurs in the 5′ to 3′ direction; one strand is rapidly copied (the leading strand) while the other (the lagging strand) is discontinuously copied as small fragments (Okazaki pieces) that are enzymatically linked by way of ligases and DNA polymerases. As the circular chromosome unwinds, topoisomerases, or DNA gyrases, function to relax the supercoiling that occurs. Finally, termination and segregation of newly replicated genetic material takes place, linked to cellular division, so that each daughter cell obtains a full complement of genetic material.

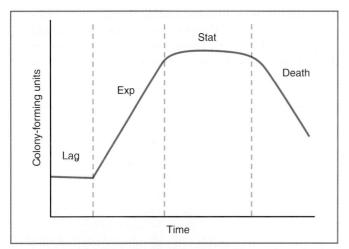

Figure 11-6. Bacterial growth represented by the number of colony-forming units versus time. Growth phases depend on environmental conditions. During the lag phase (*Lag*), bacteria adapt to growth conditions; individual bacteria are maturing but not yet able to divide. In the exponential phase (*Exp*), organisms are reproducing at their maximum rate. The growth rate slows during the stationary phase (*Stat*) owing to depletion of nutrients and exhaustion of available resources. Finally, in the death phase (*Death*), bacteria run out of nutrients and die.

GENETICS

Histones and Chromosomes

Human genetic material is complexed with histone proteins (two each of H2a, H2b, H3, and H4, and one linker H1 molecule) and organized into nucleosomes, which are further condensed into chromatin. This gives rise to the chromosome structure. Approximately 3 billion base pairs of DNA encoding 30,000 to 40,000 genes are present within the 23 pairs of tightly coiled chromosomes.

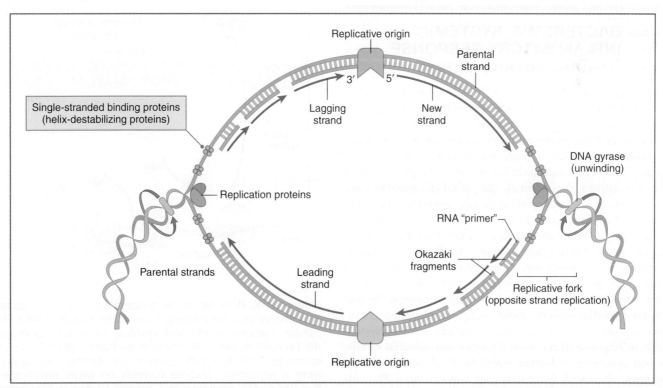

Figure 11-7. Bacterial replication as a three-phase process. Replication begins with unwinding of the DNA beginning at a unique sequence termed the origin of replication. Gyrases (topoisomerases) unwind the chromosome, while single-stranded binding proteins hold open the double helix to allow polymerases to copy the strands through addition of complementary nucleotides in a 5′ to 3′ direction. Bidirectional copying occurs by synthesizing short Okazaki fragments on the lagging strand, which are later connected by specific ligases.

Gene Expression and Regulation

Bacteria lack nuclear membranes, allowing simultaneous transcription of DNA to messenger RNA (mRNA) and translation of proteins. Although bacterial mRNA is short lived, each message may be translated approximately 20 times before degradation by nucleases. Messenger RNA is polycistronic, containing genetic information to translate more than one protein. An operon is a group of genes that includes an operator and a common promoter region plus one or more structural genes. Transcriptional regulation occurs through inducer or repressor proteins that interact with structural regions (physical sequences) of the operon to regulate the rate of protein synthesis. Multiple ribosomal units are present on each mRNA, allowing for large numbers of proteins to be produced prior to mRNA destruction. Translation of mRNA usually begins at an AUG start codon preceded by a specific ribosome binding sequence (called the *Shine-Dalgarno region*).

KEY POINTS ABOUT BACTERIAL GENETICS

- Bacterial genes are located within the cytoplasm on a supercoiled chromosome as well as on extrachromosomal plasmids.
- Many antibiotic resistance genes are located on plasmids.
- Transfer of genetic material between species may occur through various mechanisms, including conjugation, transduction, and transformation.
- Gene expression and protein synthesis are under tight regulatory control.

BACTEREMIA, SYSTEMIC INFLAMMATORY RESPONSE SYNDROME, AND SEPSIS

The majority of infections are self-limiting or subclinical in nature with only minimal or localized inflammatory responses evident due to microbial invasion. Symptoms can be transient, or, if bacterial agents persist, they can cause clinical symptoms of higher order, such as those seen in rhinitis and sinusitis, nephritis, or even endocarditis. Once bacteria are present in the bloodstream (bacteremia), the pathologic outcomes are more severe. Systemic inflammatory response syndrome can serve as a precursor to full-blown sepsis, in which profound global immune responses affect host function. In severe septic states, organ perfusion is affected, leading to hypoxia and hypotension. Changes in mental status also occur. The pathogenesis of sepsis is very complex, and is dependent in part on the individual organism causing the syndrome. Proinflammatory cytokines (e.g., interleukin-6 and tumor necrosis factor-α) are released by innate immune system cells in response to recognized factors and bacterial motifs, which synergize to further stimulate T-cell and B-cell responses, often with tissue-damaging consequences. Platelet-activating factor, leukotrienes, and prostaglandins are released, along with other bioactive metabolites of the arachidonic pathway, priming additional granulocytes to release toxic oxidative radicals. Septic shock eventually ensues, leading to outcomes of multiple organ failure and poor prognosis.

BACTERIAL TOXINS: VIRULENCE FACTORS THAT TRIGGER PATHOLOGY

Bacterial toxins are biologic virulence factors that prepare the host for colonization. By definition, a toxin triggers a destructive process (Fig. 11-8). Toxins can function in multiple ways, for example, by inhibiting protein synthesis (diphtheria toxin),

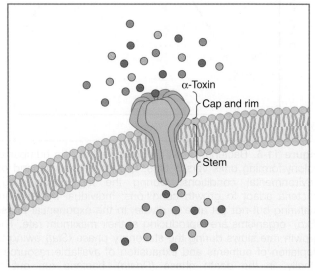

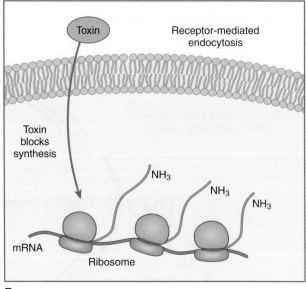

Figure 11-8. Bacterial toxins function as virulence factors. Two mechanisms for bacterial toxin action include damage to cellular membranes (**A**) and inhibition of protein synthesis (**B**). Damage to cellular membranes, such as by *Staphylococcus aureus* or *Clostridium perfringens* α toxin, functions by assembling a heptomeric prepore complex on target membranes that undergoes conformational change to disrupt membrane permeability and affect influx and efflux of ions. Inhibition of protein synthesis, as exemplified by *Shigella dysenteriae* Shiga toxin, *Escherichia coli* heat-labile toxin I, and cholera and pertussis toxins, which work as substrates for elongation factors and ribosomal RNA.

activating second messenger pathways (*Bacillus anthracis* edema factor or cholera toxin), activating immune responses (*S. aureus* superantigens), damaging cell membranes (*E. coli* hemolysin), or by general action of metalloprotease activity (*Clostridium tetani* tetanus toxin). Toxins come in a variety of forms. LPS is considered a powerful endotoxin; its activity has been attributed to the lipid A portion of the molecule. Indeed, septic shock is thought to be caused by LPS induction of proinflammatory mediators. In contrast to bound endotoxins, bacterial exotoxins are soluble mediators located in the bacterial cytoplasm or periplasm that are either excreted or released during bacterial cell lysis or destruction. A specific class of exotoxins, called enterotoxins, is toxic to the intestinal tract, causing vomiting and diarrhea. Well-defined toxins (such as enterotoxins, neurotoxins, leukocidins, and hemolysins) are classified in terms of the specific target cell or site affected. Table 11-1 lists major toxins and their mechanism of action.

TABLE 11-1. Important Bacterial Toxins and Their Mechanism of Action

TOXIN	ORGANISM	MECHANISM	CLINICAL FEATURES
Anthrax toxins (edema toxin [EF], lethal toxin [LF])	*Bacillus anthracis*	Adenylyl cyclase (EF), metalloprotease (LF)	Edema and skin necrosis; shock
Adenylate cyclase toxin	*Bordetella pertussis*	Adenylyl cyclase	Tracheobronchitis
Botulinum toxins (C2/C3 toxin)	*Clostridium botulinum*	Blocks release of acetylcholine, ADP-ribosyltransferase	Muscle paralysis, botulism
Toxin A/toxin B	*Clostridium difficile*	Inhibits cytoskeletal action in epithelial cells	Diarrhea, vomiting
Lecithinase (α-toxin; perfringolysin O)	*Clostridium perfringens*	Phospholipase	Gangrene; destruction of phagocytes
Tetanus toxin	*Clostridium tetani*	Blocks release of inhibitory neurotransmitters	Spasms and rigidity of the voluntary muscles; characteristic symptom of "lockjaw"
Diphtheria toxin	*Corynebacterium diphtheriae*	ADP-ribosylates EF-2, inhibiting protein synthesis	Respiratory infection; complicating myocarditis with accompanying neurologic toxicity
CNF-1, CNF-2	*Escherichia coli*	Affects ρ-GTP–binding regulators	Diarrhea
Heat-stable toxin	*Escherichia coli*	Secondary message regulation	Diarrhea
Hemolysin	*Escherichia coli*	Heptameric pore-forming complex (hemolysin)	Urinary tract infections
Shiga-like toxin	*Escherichia coli*	Stops host protein synthesis	Hemolytic-uremic syndrome, dysentery
Exotoxin A	*Pseudomonas aeruginosa*	ADP-ribosylates elongation factor-2 (EF-2), inhibiting protein synthesis	Respiratory distress; possible role as virulence factor in lung infections of cystic fibrosis patients
Shiga toxin	*Shigella dysenteriae*	Stops host protein synthesis	Dysentery
α-Toxin	*Staphylococcus aureus*	Heptameric pore-forming complex (hemolysin)	Abscess formation
Toxic shock syndrome toxin 1 (TSST-1)	*Staphylococcus aureus*	Superantigen activates T-cell populations, cross-linking V_β TCR and class II MHC	Cytokine cascade elicits shock; capillary leak syndrome and hypotension
Pneumolysin	*Streptococcus pneumoniae*	Pore-forming complex (hemolysin)	Pneumonia
Pyrogenic exotoxin	*Streptococcus pyogenes*	Superantigen activates T-cell populations, cross-linking V_β TCR and class II MHC	Cytokine cascade elicits shock; capillary leak syndrome and hypotension
Streptolysin O	*Streptococcus pyogenes*	Pore-forming complex (hemolysin)	"Strep" throat, scarlet fever
Cholera toxin	*Vibrio cholerae*	Disrupts adenylyl cyclase	Watery diarrhea, loss of electrolytes and fluids

ADP, adenosine diphosphate; *EF*, elongation factor; *LF*, lethal factor; *GTP*, guanosine triphosphate; *TCR*, T-cell receptor; *MHC*, major histocompatability complex.

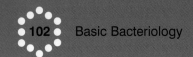

KEY POINTS ABOUT BACTERIAL TOXINS

- Many bacteria synthesize toxins that serve as primary virulence factors, inducing pathologic damage to host tissue.
- Toxins may function to establish productive colonization conditions and work by damaging host cell membranes, by inhibiting host cell protein synthesis, and by activating secondary messengers that adversely affect host cell function.

CLINICAL DIAGNOSIS

It is critical to adequately prepare clinical specimens for purposes of organism identification. A complete detailing of diagnostic parameters is beyond the scope of this text; however, it is important to mention a number of classical techniques commonly used in the clinical laboratory. In most cases, isolation of organisms may be accomplished using culture methods in defined medium, which also allows for determination of antibiotic susceptibility. Growth on blood agar can determine evidence of hemolytic colonies, such as is seen with β-hemolytic streptococci. Organisms may be detected via visualization using specific stains and matching morphological characteristics. For example, a gram-negative reaction represents an organism with a cell envelope that has an outer membrane with only a thin peptidoglycan layer; a gram-positive reaction is indicative of a cell envelope with a thick peptidoglycan cell wall and no outer membrane. In contrast, an acid-fast cell envelope is one that has a thick peptidoglycan layer similar to gram-positive bacteria, but which contains a high concentration of waxy, long-chain fatty acids (mycolic acids), as seen with the mycobacterial species. And biochemical tests, such as those for catalase and coagulase, are important markers for organisms known to disrupt tissues and clot vessels during infection.

Other methods employ molecular techniques, such as the use of polymerase chain reaction to amplify specific and unique nucleotide sequences. Antibody-based methodologies, such as enzyme-linked immunoassay and agglutination technologies, can detect species-specific and serovar-specific antigens. Finally, detection of antibodies can also indicate the presence of organisms in the host, with antibody isotype identification indicative of present or past infection.

MAJOR ANTIMICROBIAL AGENTS AGAINST BACTERIA

Antimicrobial agents can be categorized as molecules that act to kill or inhibit bacterial growth by interfering with (1) cell wall synthesis, (2) ribosomal function and protein synthesis, (3) nucleic acid synthesis, (4) folate synthesis, or (5) plasma membrane integrity. Many useful antimicrobial agents and their mechanisms of action are listed in Table 11-2. In brief, cell wall synthesis is inhibited by β-lactams, such as penicillins and cephalosporins, which inhibit peptidoglycan polymerization. In addition, vancomycin inhibits synthesis of cell wall substrates. Aminoglycosides, streptomycin, tetracycline, chloramphenicol, erythromycin and related macrolides (clarithromycin, azithromycin), and clindamycin all interfere with ribosome function through binding to the 30 S or 50 S ribosomal subunit. Quinolones bind to a bacterial complex of DNA and DNA gyrase, blocking DNA replication. Rifampin blocks transcription of mRNA synthesis by binding and inhibiting RNA polymerase. Nitroimidazoles damage DNA. Nalidixic acids (floxacin, ciprofloxacin, norfloxacin) inhibit DNA unwinding needed for DNA synthesis. Sulfonamides and

TABLE 11-2. Selected Antimicrobial Agents and General Mechanisms of Action

CELL WALL SYNTHESIS INHIBITION	PROTEIN SYNTHESIS INHIBITION	NUCLEIC ACID FUNCTION INHIBITION	FOLIC ACID SYNTHESIS INHIBITION	DAMAGE TO CELL MEMBRANE
Penicillins	Aminoglycosides	Quinolones	Trimethoprim	Polymyxins
Penicillin G and V	Streptomycin	Sulfanilamide	Sulfanilamide	Bacitracin
Ampicillin	Neomycin	Rifampicin	Trimethoprim-	Amphotericin
Amoxycillan	Amikacin	Nalidixic acids	sulfamethoxazole	Nystatin
Carbenicillin	Kanamycin	Floxacin	(cotrimoxazole)	Imidazole
Methicillin	Gentamicin	Ciprofloxacin		Ketoconazole
Cephalosporins	Tobramycin	Norfloxacin		
Cefamandole	Tetracycline	Levofloxacin		
Vancomycin	Chlortetracycline	Moxifloxacin		
Bacitracin	Oxytetracycline	Gatifloxacin		
Novobiocin	Doxycycline	Metronidazole		
Cycloserine	Minocycline			
	Erythromycin			
	Clarithromycin			
	Azithromycin			
	Clindamycin			
	Spectinomycin			
	Chloramphenicol			
	Lincosamides			
	Cycloserine			
	Streptpgramin			

trimethoprim block the synthesis of folate needed for DNA replication. Polymyxins and amphotericin disrupt the plasma membrane, causing leakage. The plasma membrane sterols of fungi are attacked by polyenes (amphotericin) and imidazoles.

Bacterial resistance is a natural outcome of evolution and environmental pressure. Resistance factors can be encoded on plasmids or within the bacterial chromosome. The etiology of antimicrobial resistance may involve mechanisms that limit entry of the drug, changes in the receptor (target) of the drug, or metabolic inactivation of the drug. Bacteria acquire genes conferring antimicrobial resistance in numerous ways including spontaneous DNA mutation, bacterial transformation, and plasmid transfer.

PHARMACOLOGY

β-Lactam Antibiotics

β-Lactam antibiotics are inhibitors of cell wall synthesis, working to limit transpeptidase and carboxypeptidase action involved in terminal cross-linking of glycopeptides in the formation of the peptidoglycan layer. Penicillins derived from the mold *Penicillium* consist of a β-lactam ring coupled to a thiazolidine ring. Addition of defined side chains to the free amino group produces a range of synthetic antibiotics including ampicillin and methicillin, which are active against both gram-negative and gram-positive species.

KEY POINTS ABOUT ANTIMICROBIAL AGENTS

- Antimicrobial agents function through inhibition of synthetic pathways required for bacterial growth and via direct damage to bacterial membranes.
- Bacterial antibiotic resistance is a natural phenomenon that may occur by natural mutations to existing genes or through additions to nucleic acid content via transformation or acquisition of plasmid DNA from other bacteria.

KEY POINTS

- Bacteria are classified according to morphologic structure, metabolic activity, and environmental factors needed for survival. Gram-positive organisms have a thick peptidoglycan outer layer with teichoic and lipoteichoic acid present, while the gram-negative bacteria have a thin peptidoglycan component surrounding a periplasmic space, and an outer membrane with lipoproteins.
- Bacterial gene expression and protein production are under tight regulatory control, with genes present in a double-stranded, supercoiled DNA chromosome or on plasmids. Genetic material is readily passed between organisms, by molecular mechanisms that include conjugation, transduction, and transformation.
- During sepsis, cytokines released by innate immune cells in response to recognized bacterial motifs trigger T-cell and B-cell responses, often with tissue-damaging consequences.
- Bacterial toxins serve as virulence factors, inducing pathologic damage to host tissue and assisting in avoidance of immune surveillance.
- Antimicrobial therapeutics function to limit growth by inhibition of biochemical pathways of cell wall synthesis, ribosomal function, nucleic acid synthesis, and energy production. Resistance through mutations, by insertion of nucleic acids, or via nuclear acquisition allows bacteria to evade drug-related metabolic inactivation.

Self-assessment questions can be accessed at www.StudentConsult.com.

Clinical Bacteriology 12

CONTENTS

GRAM-POSITIVE COCCI
  Staphylococci
  Streptococci and Enterococci
  Other Gram-Positive Cocci of Medical Importance
GRAM-NEGATIVE COCCI
  Neisseria
  Veillonella
AEROBIC GRAM-POSITIVE BACILLI
  Bacillus
  Lactobacillus
  Listeria
  Erysipelothrix
  Corynebacteria
ANAEROBIC GRAM-POSITIVE BACILLI
  Clostridium
  Actinomyces
  Bifidobacterium
  Eubacterium and Propionibacterium
AEROBIC GRAM-NEGATIVE BACILLI
  Enterobacteriaceae
  Other Pathogenic Enterobacteriaceae
  Haemophilus
  Legionella
  Bordetella
ORGANISMS OF ZOONOTIC ORIGIN
  Pasteurella
  Brucella
  Francisella
  Bartonella
  Vibrio
CAMPYLOBACTER* AND *HELICOBACTER
NONFERMENTERS
  Pseudomonas
  Acinetobacter
ANAEROBIC GRAM-NEGATIVE BACILLI
  Bacteroides
  Fusobacterium
SPIROCHETES
  Treponema
  Borrelia
  Leptospira
OBLIGATE INTRACELLULAR PATHOGENS
  Chlamydia
  Rickettsia, Ehrlichia, and Coxiella
ACID-FAST ORGANISMS
  Mycobacterium
  Nocardia
  Acid-Fast Intestinal Coccidia
MYCOPLASMA

A review of clinically important bacteria includes a wide range of organisms, representing pathogens that cause disease with complex etiologies. More often than not, the basis of pathogenicity depends on virulence factors produced by the bacteria that mediate environmental conditions and affect host immune function. Opportunistic infections represent organisms that elicit pathogenic responses in immunocompromised or immunosuppressed hosts; in many cases, these bacteria represent normal flora that otherwise rarely cause disease. Table 12-1 outlines the major pathogens of medical significance.

●●● GRAM-POSITIVE COCCI

The *Staphylococcus*, *Streptococcus*, and *Enterococcus* spp. are nonmotile, non–spore-forming, gram-positive organisms that cause pyogenic (producing pus) and pyrogenic (producing fever) infections. They represent a major population responsible for cutaneous infections and are causative agents for pathology manifested as systemic disease. Typical skin lesions associated with pyogenic gram-positive organisms include abscesses with central necrosis and pus formation. These include boils, furuncles, and impetigo (cutaneous, pustular eruptions), carbuncles (subcutaneous) and erysipelas (deep red, diffuse inflammation), paronychias (nailbed infection), and styes (eyelid infection). Systemic infections include bacteremia, food poisoning, endocarditis, toxic shock syndrome, arthritis, and osteomyelitis. The pathogenic mechanisms for clinical disease occur through a combination of toxins and enzymes produced by organisms and relative effects on immune cells involved in combating localized infections. For example, staphylococci produce multiple virulence factors, including exotoxins that regulate cytokines, leukocidins that kill polymorphonuclear cells, and α-toxins (hemolysins) that contribute to local tissue destruction and lysing of red blood cells (Fig. 12-1). They also possess a catalase that inactivates hydrogen peroxide, a key component released by neutrophils responding to infection and found within lysosomes of activated macrophages.

Staphylococci

The staphylococci are facultative anaerobes, morphologically occurring in grape-like clusters. They are major components of the normal flora of skin and nose and are catalase positive. *Staphylococcus aureus* (coagulase-positive) is one of the most common causes of opportunistic infections in the hospital and

TABLE 12-1. Major Bacterial Pathogens of Medical Interest

BACTERIAL TYPE	DESCRIPTION	EXAMPLE
Gram-positive cocci	Fermentative Catalase positive	*Staphylococcus*
	Catalase negative	*Streptococcus* *Enterococcus*
	Oxidative	*Micrococcus*
Gram-negative cocci	Aerobic, non–spore forming	*N. gonorrhoeae*, *N. meningitidis*, *Veillonella*
Gram-positive rods	Endospore forming	*Bacillus* *Clostridium*
	Regular, non–endospore forming	*Lactobacillus* *Listeria* *Erysipelothrix*
	Irregular, non–endospore forming	*Corynebacterium*
Aerobic gram-negative rods	Enterobacteriaceae	*E. coli* *Shigella* *Salmonella* *Yersinia* *Edwardsiella* *Citrobacter* *Klebsiella* *Enterobacter* *Serratia* *Proteus* *Morganella* *Providencia*
	Respiratory tract	*H. influenzae* *L. pneumophila* *B. pertussis*
	Zoonotic origin	*P. multocida* *Brucella* spp. *F. tularensis*
	Nonfermentative	*Pseudomonas* *Acinetobacter*
	Miscellaneous gram-negative rods	*V. cholerae* *C. jejuni* *H. pylori*
Anaerobic gram-negative rods	Digestive tract	*Bacteroides*
	Pleuropulmonary	*Fusobacterium*
Spirochetes	Spirochaetaceae	*Treponema* (syphilis) *Borrelia* (Lyme disease) *B. recurrentis*, *B. hermsii* (relapsing fever, antigenic change) *Leptospira* ssp.
Obligate intracellular pathogens	Atypical cell wall	*Chlamydia* *Rickettsia* *Ehrlichia* *Coxiella*
Acid-fast organisms	Mycolic acid cell wall	*M. tuberculosis*, *M. avium*, *M. leprae*
	No mycolic acids in cell wall	*Nocardia*
	Intestinal coccidia	*Cryptosporidium*, *Cyclospora*, *Isospora*
Other pathogens	No cell wall	*Mycoplasma*

community, including pneumonia, osteomyelitis, septic arthritis, bacteremia, endocarditis, and skin infections. In addition, ingested food contaminated with *S. aureus* enterotoxin can readily lead to vomiting, nausea, diarrhea, and abdominal pain. *S. aureus* produces TSST-1, a superantigen that has been directly linked to toxic shock syndrome. The superantigen is immune deceptive, able to activate a large percentage of nonspecific T cells via exogenous cross-linking of the T cell

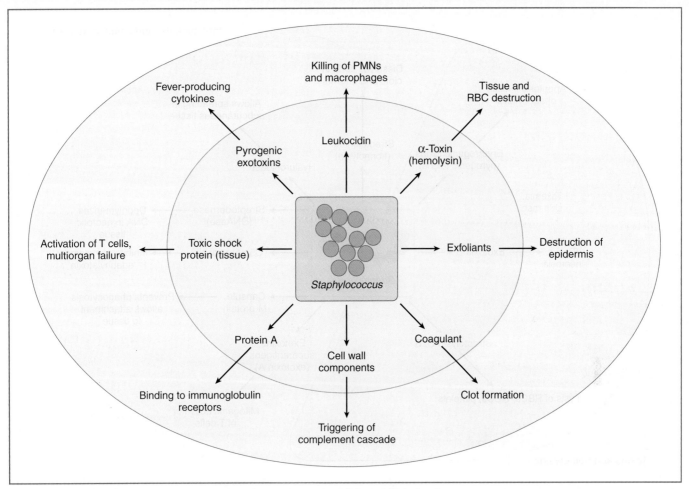

Figure 12-1. Pathogenic mechanisms of *Staphylococcus* spp. *RBC*, red blood cell.

receptor with major histocompatability complex molecules on antigen-presenting cells. *S. aureus* also produces an exfoliative toxin that causes scalded skin syndrome in babies. *S. aureus* produces various exotoxins, as well as tissue-degrading enzymes involved in disease spreading (lipase and hyaluronidase), and protein A, which binds to the Fc portion of immunoglobulin (Ig) G, thus inhibiting induction of phagocytosis by polymorphonuclear cells and macrophages and induction of complement cascades.

S. epidermidis is a less common cause of opportunistic infection than *S. aureus* and is relevant as a mediator of nosocomial infections. *S. epidermidis* is also a major component of the skin flora and mucous membranes, can easily be cultured from wounds and blood, and is commonly found on catheter tips. A closely related staphylococcal species, *S. saprophyticus*, is a major cause of urinary tract infections in young women. Both *S. epidermidis* and *S. saprophyticus* are coagulase negative.

Streptococci and Enterococci

The *Streptococcus* spp. are subdivided into four groups with overlapping ability to cause clinical disease ranging from pharyngitis and general cellulitis to toxic shock syndrome and severe sepsis. The streptococci of medical importance may be identified according to their hemolytic patterns or according to antigenic differences in carbohydrates located within their cell wall. All streptococci are catalase negative and exhibit hemolysins of type α or β (streptolysin O and streptolysin S). Group A streptococci (*S. pyogenes*) is the most clinically important member of the *Streptococcus* spp.; *S. pyogenes* is the causative agent of pharyngeal infection, acute rheumatic fever (nonsuppurative disease of the heart and joints), and glomerulonephritis. In addition, it is the etiologic agent of scarlet fever, with erythrogenic (pyrogenic) toxins causing characteristic rash. One of the pyrogenic toxins is a superantigen, causing mitogenic T-cell response in a non–antigen-specific mediated manner. Other toxins (pyrogenic toxins A, B, and C) when released result in severe edema and necrotizing myositis and fasciitis. Some of the pathogenic mechanisms are depicted in Figure 12-2. Finally, it is hypothesized that acute rheumatic fever and subsequent inflammatory lesions of the joints and heart are resultant autoimmune dysfunction derived from molecular mimicry against antigens derived from group A β-hemolytic streptococcal agents.

S. agalactiae (group B streptococcus) readily colonizes the vaginal region and is a common cause of neonatal bacteremia and sepsis, pneumonia, and meningitis due to transmission from mother to child before or after childbirth. The group D streptococci include *S. bovis* and the enterococci. *Enterococcus faecalis* (previously identified as *Streptococcus faecalis*) is a causative agent of urinary and biliary tract infections and

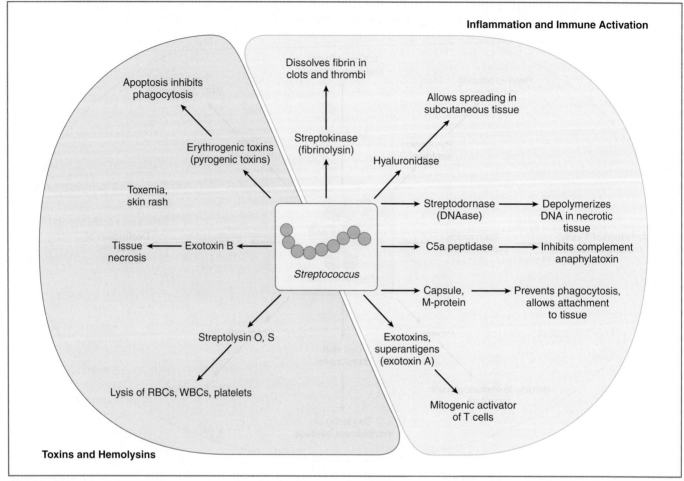

Figure 12-2. Pathogenic mechanisms for group A streptococci (*Streptococcus pyogenes*). *RBCs*, red blood cells; *WBCs*, white blood cells.

also contributes to bacteremia and endocarditis. *E. faecalis* is γ-hemolytic and has been linked to colon carcinomas. The bacterium *S. viridans* is responsible for approximately half of all cases of bacterial endocarditis. Members of this group include *S. mutans*, *S. sanguis*, and *S. salivarius*. Although these organisms are normally found as oral bacterial flora, entry into the bloodstream can lead to fever and embolic events. Group C streptococci (*S. equisimilis*, *S. zooepidemicus*, *S. equi*, *S. dysgalactiae*) primarily cause diseases of animals and pose little threat to immunocompetent hosts. Likewise, groups E, F, G, H, and K to U species rarely cause pathogenic disease.

S. pneumoniae (referred to as pneumococci) is a leading cause of pneumonia, often with onset after damage to the upper respiratory tract (e.g., following viral infection). Although *S. pneumoniae* is hemolytic, there is no group antigen and there are no main exotoxins that contribute to pathogenesis. The organism often spreads, causing bacteremia and meningitis, and may also cause middle ear infections (otitis media). *S. pneumoniae* has an antiphagocytic capsule (antigenically effective as a vaccine target) and produces a pneumolysin that degrades red blood cells to allow productive spread from respiratory membranes to the blood. It also produces an IgA protease that more readily allows colonization of respiratory mucosa. Complement activation by teichoic acid may explain the attraction of large numbers of inflammatory cells to the focal site of infection.

Other Gram-Positive Cocci of Medical Importance

The *Micrococcus* spp. include organisms that may produce pathology in immunocompromised individuals (those with neutropenia, severe combined immunodeficiency, or acquired immunodeficiency). Of these, *Stomatococcus mucilaginosus*, normally a soil-residing organism, may induce disease. *Peptostreptococcus* is an anaerobic counterpart of *Streptococcus*. Peptostreptococci are small bacteria that grow in chains; are usually nonpathogenic; and are found as normal flora of the skin, urethra, and urogenital tract. Under opportunistic conditions, they can infect bones, joints, and soft tissue. *Peptostreptococcus magnus* is the species most often isolated from infected tissues.

●●● GRAM-NEGATIVE COCCI

Neisseria

The *Neisseria* genus consists of aerobic, non–spore-forming gram-negative diplococcobacilli that reside in mucous membranes. They are nonmotile, oxidase-positive, glucose-fermenting microbes that require a moist environment and warm temperatures to achieve optimum growth. The two most clinically significant members of the genus *Neisseria* are *N. gonorrhoeae* (gonococcus) and *N. meningitidis* (meningococcus). Infection by *N. gonorrhoeae* is referred to as a *gonococcal infection* and is transmitted by intimate contact with the mucous membranes. In infected males, the disease is characterized by urethritis with a urethral pus discharge; if left untreated, resulting complications such as prostatitis and periurethral abscess may occur. Females with gonorrhea exhibit vaginal discharge (cervicitis or vulvovaginitis) with accompanying abdominal pain and nonmenstrual bleeding. As with most other sexually transmitted diseases, gonorrhea is prevalent in young adult and homosexual populations. *N. gonorrhoeae* is sensitive to antibiotics; the common association with chlamydial infection dictates using a therapeutic regimen of cephalosporin (ceftriaxone) and a tetracycline or quinolone to kill organisms. If left untreated, *N. gonorrhoeae* can cause meningitis with septicemia and resulting arthritis and acute endocarditis upon further dissemination of organisms.

N. meningitidis colonizes the nasopharynx and is the second most prevalent causative agent of meningitis in the United States. Upon invasion of blood, it may cause purpura, endotoxic shock, and meningitis with characteristic inflammation of membranes covering the central nervous system (CNS). Early symptoms are headache, fever, and vomiting; death can quickly follow owing to focal cerebral involvement from the highly toxic lipopolysaccharide. Antibody-dependent complement-mediated killing is a critical component of host defenses against the meningococci. *N. meningitidis* also has an antiphagocytic capsule, which contributes to its virulence. Different strains of *N. meningitidis* are classified by their capsular polysaccharides, with nine divisible serogroups (A, B, C, D, X, Y, Z, W135, and 29E). *N. meningitidis*, as well as *N. gonorrhoeae*, produces proteases that target IgA to promote virulence. Organisms can assume carrier status, with subsequent disease developing only in a few carriers. Most infected patients can be treated with penicillin G, while rifampin may be used prophylactically to prevent reactivation of disease.

Veillonella

Veillonella spp. are nonmotile, gram-negative diplococci that are the anaerobic counterpart of *Neisseria*. *Veillonella* is part of the normal flora of the mouth and gastrointestinal tract and may be found in the vagina as well. Although of limited pathogenicity, *Veillonella* is often mistaken for the more serious gonococcal infection. *Veillonella* spp. are often regarded as contaminants; they are often associated with oral infections; bite wounds; head, neck, and various soft tissue infections; and they have also been implicated as pathogens in infections of the sinuses, lungs, heart, bone, and CNS. Recent reports have also indicated their isolation in pure culture in septic arthritis and meningitis.

PATHOLOGY

Pus and Abscess Formation

An accumulation of pus in an enclosed tissue space is known as an *abscess*. Pus, a whitish-yellow substance, is found in regions of bacterial infection including superficial infections such as pimples. Pus consists of macrophages and neutrophils, bacterial debris, dead and dying cells, and necrotic tissue. Necrosis is caused by released lysosomes, including lipases, carbohydrases, proteases, and nucleases.

●●● AEROBIC GRAM-POSITIVE BACILLI

Bacillus

Bacillus is a genus of gram-positive bacteria that are ever present in soil, water, and airborne dust. *Bacillus* may be found as a natural flora in the intestines. *Bacillus* has the ability to produce endospores under stressful environmental conditions. The organism is nonmotile and nonhemolytic and is highly pathogenic. The only other known spore-producing bacterium is *Clostridium*. Although most species of *Bacillus* are harmless saprophytes, two species are considered medically significant: *B. anthracis* and *B. cereus*. *B. anthracis* is a nonhemolytic, nonmotile, catalase-positive bacterium that causes anthrax in cows, sheep, and sometimes humans. Under the microscope, *B. anthracis* cells appear to have square ends and seem to be attached by a joint to other cells. Anthrax is transmitted to humans by cutaneous contact (infection of abrasions) with endospores, or more rarely by inhalation. Rare cases of gastrointestinal infection may occur. Cutaneous anthrax causes ulceration, with a distinctive black necrotic center surrounded by an edematous areola with pustules. Pulmonary and gastrointestinal infections are more likely to result in toxemia. *B. anthracis* secretes three toxins to help evade host immune response through exertion of apoptotic effect on responding cells (Fig. 12-3). Two of these toxins are edema factor (EF) and lethal factor (LF), both of which have negative effects via enzymatic modification of substrates within the cytosol of the host cell. Protective antigen, the third component, binds to a cellular receptor, termed *anthrax toxin receptor*, and functions in transporting both EF and LF into host cells.

Unlike *B. anthracis*, *B. cereus* is a motile, catalase-positive bacterium that is the causative agent of a toxin-mediated food poisoning. It is a common soil and water saprophyte that, upon ingestion, releases two toxins into the gastrointestinal tract that cause vomiting and diarrhea; the clinical manifestations are similar to those of *Staphylococcus* food poisoning. One of the endotoxins is similar to the heat-labile toxin of *Escherichia coli*, with associated activation of cyclic adenosine monophosphate (cAMP)-dependent protein kinase activity in enterocytes underlying watery diarrhea production.

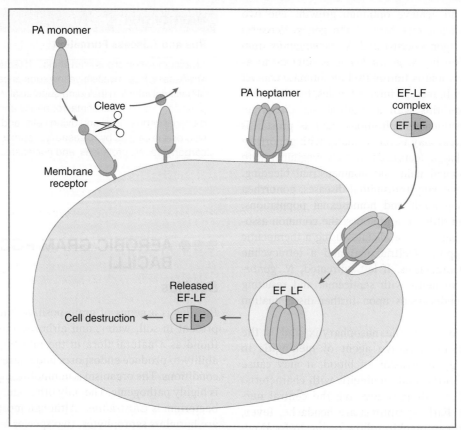

Figure 12-3. Mechanisms of anthrax toxin. Assembly of lethal *Bacillus anthracis* toxin on the cell surface causes anthrax poisoning. The bacteria produce three proteins which, when combined, kill cells. The proteins—protective antigen (*PA*; monomers forming a heptamer prepore), lethal factor (*LF*; primary protease-cleaving mitogen-activated protein kinase), and edema factor (*EF*; adenylate cyclase, converting adenosine triphosphate into cyclic adenosine monophosphate)—are not toxic as monomers released by the bacteria, but they cause cell disruption upon interacting.

Proper cold storage of food is recommended immediately after preparation to limit growth and toxin production.

Lactobacillus

Lactobacillus is a gram-positive, facultatively anaerobic, non-motile, non–spore-forming bacterium that ferments glucose into lactose, thus earning its name. The most common application of *Lactobacillus* is for dairy production. This genus contains several species that belong to the natural flora of the vagina, with other related organisms found in the colon and mouth. *Lactobacillus* derives lactic acid from glucose, creating an acidic environment that inhibits growth of other bacterial species, which contributes to urogenital infections. Infection may occur in the mouth, in which case, colonization may affect dentition; if the infection in enamel goes unchecked, acid dissolution can advance cavitation extending through the dentin (the component of the tooth located under the enamel) to the pulp tissue, which is rich in nerves and blood vessels. In rare cases of infection, treatment usually consists of high doses of penicillin in combination with gentamicin.

PATHOLOGY

Pathophysiology of Diarrhea

The causes of diarrhea may be identified as defects in absorption, secretion, or motility. Factors that alter the normal transit of a meal through the alimentary canal affect the consistency of the fecal contents; increases in motility are correlated with diarrhea. Infection that alters the function of the enterocytes of the small intestine leads to massive water flux to the colon. Specific effectors that increase cAMP levels in enterocytes stimulate secretion of chloride and bicarbonate, with associated sodium and water secretion.

Listeria

Listeria is a gram-positive, catalase-positive rod (diphtheroid) that is not capable of forming endospores. Two species are of human pathogenic significance: *L. monocytogenes* and *L. ivanovii*. In particular, *L. monocytogenes* causes meningitis and sepsis in newborns and accounts for 10% of community-acquired bacterial meningitis in adults. While host monocytes are critical for control and containment of *Listeria*, they also are involved in disseminating infection to other areas of the body. *Listeria* is also diarrheagenic in humans, with those infected having vomiting, nausea, and diarrhea. Ingestion of *Listeria* from unpasteurized milk products can lead to bacteremia and septicemia with meningoencephalitis. When transmitted across the placenta to the fetus, infection can lead to placentitis, neonatal septicemia, and possible abortion. Individuals at particular risk for listeriosis include newborns, pregnant women and their fetuses, the elderly, and persons lacking a healthy immune system. The bacterium usually causes septicemia and meningitis in patients with suppressed immune function. Antibiotics are recommended for treatment of infection because most strains of *Listeria* are sensitive to ampicillin plus an aminoglycoside. Identification uses β-hemolysis on blood agar plates.

Erysipelothrix

Erysipelothrix rhusiopathiae is a common veterinary pathogen; however, infection of human hosts occurs. In humans, *Erysipelothrix* is an aerobic, non–spore-forming, gram-positive bacillus that has been linked to skin infections in meat and fish handlers; the most common presentation is cellulitis (erysipeloid), a localized cutaneous infection. A more serious condition may occur involving lesions that progress from the initial site of infection or appear in remote areas. A severe form of disease is a septicemia that is almost always linked to endocarditis. Treatment usually consists of penicillin G, ampicillin, cephalothin, or other β-lactam antibiotics. Most clinical strains are resistant to vancomycin.

Corynebacteria

The coryneform group includes several genera of non–spore-forming rods that are ubiquitous in nature. They are gram-positive bacteria that include the clinically important *Actinomyces* and *Corynebacterium*. *Corynebacterium* spp. are nonmotile, facultatively anaerobic bacteria that are usually saprophytic and cause little harm to humans. However, *C. diphtheriae* can be pathogenic, producing the toxin that causes diphtheria, a disease of the upper respiratory system. Although other species of *Corynebacterium* can inhabit the mucous membrane, *C. diphtheria* is unique in its exotoxin formation. Pathogenesis manifests as inflammatory exudates that may spread infection to the postnasal cavity or the larynx and cause respiratory obstruction. Bacilli do not penetrate deeply into underlying tissues; rather, a powerful exotoxin is produced that has a special affinity for heart, muscle, nerve endings, and the adrenal glands. The diphtheria toxin is a heat-stable polypeptide composed of two fragments, which together inhibit polypeptide chain elongation at the ribosome. Inhibition of protein synthesis is probably responsible for both necrotic and neurotoxic effects. Patients have malaise, fatigue, fever, and sore throat; infection manifesting as anterior nasal diphtheria presents with a thick nasal discharge. *C. diphtheriae* is sensitive to penicillin, tetracycline, rifampicin, and clindamycin. The bacteria may be viewed microscopically using the Löffler methylene blue stain. An antitoxin should be administered at the first evidence of infection and should not await laboratory confirmation.

Other medically important coryneforms include *Corynebacterium ulcerans* and *Arcanobacterium haemolyticum*, causative agents of acute pharyngeal infections; *Corynebacterium pseudotuberculosis*, involved in subacute lymphadenitis; *Corynebacterium minutissimum*, associated with infections of the stratum corneum, leading to erythrasma (scaly red patches); and *Corynebacterium jeikeium*, which has been implicated in endocarditis, neutropenia, and hematologic malignancy.

●●● ANAEROBIC GRAM-POSITIVE BACILLI

Clostridium

Clostridium spp. are gram-positive, anaerobic, spore-forming rods that are motile in their vegetative form. They are ubiquitous in soil. Physically, they appear as long thick rods with a bulge at one end. *Clostridium* grows well at body temperature; in stressful environments, the bacteria produce spores that tolerate extreme conditions. These bacteria secrete powerful exotoxins responsible for diseases including those causing tetanus, botulism, and gas gangrene. The four clinically important species are *C. tetani*, *C. botulinum*, *C. perfringens*, and *C. difficile*.

C. tetani causes tetanus (lockjaw) in humans. *C. tetani* spores germinate in an anaerobic environment to form active *C. tetani* cells, which have a drumstick-shaped appearance. Growth in dead tissue allows production and release of an exotoxin (tetanospasmin) that causes nervous system irregularities that interfere with spinal cord synaptic reflexes. The toxin blocks inhibitory mechanisms that regulate muscle contraction, leading to constant skeletal muscle contraction. Prolonged infection leads to eventual respiratory failure with a high mortality rate if left untreated. Immunization is highly effective in preventing *C. tetani* infections in both children and adults and can also function to neutralize toxin after infection.

C. botulinum produces one of the most potent neurotoxins and is the cause of deadly botulism food poisoning. Airborne *Clostridium* spores can find their way into foods that will be placed in anaerobic storage. Immediate symptoms of infection include muscular weakness with blurred vision, which develops into an afebrile neurologic disorder with characteristic descending paralysis from blocked release of acetylcholine (Fig. 12-4). Immediate treatment with antitoxin is required. Infantile botulism is much milder than the adult version; honey is a common source of spores that can germinate in the intestinal tract of children.

C. perfringens is a nonmotile, invasive pathogen responsible for gas gangrene (clostridial myositis or myonecrosis). The organism is commonly found among the gastrointestinal tract flora and can be found colonizing the skin, especially in the perirectal region. The organism can easily invade wounds that come into contact with soil. *C. perfringens* cells proliferate after spore germination occurs, releasing a variety of virulence factors that can degrade tissue. *C. perfringens* produces a lecithinase capable of lysing cells, a protease, hyaluronidase,

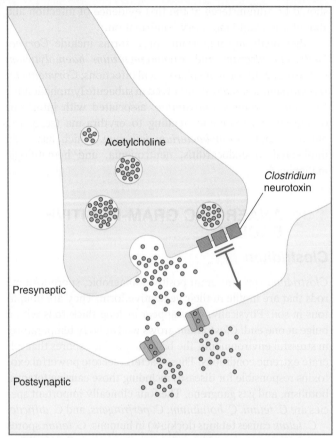

Figure 12-4. Action of clostridial neurotoxins. *Clostridium* produces an endopeptidase that blocks the release of acetylcholine at the myoneural junction. Muscle paralysis is the result. Both the botulinum toxin and the tetanus toxin interfere with vesicle formation at synaptic junctions, resulting in muscle spasms and loss of neural signal.

collagenase, and other hemolysins. The combination of virulence factors produced is strain dependent. All strains produce lecithinase (also called α-toxin), which plays a central role in pathogenesis of gas gangrene. Lecithinase can lyse white blood cells, enhancing its ability to evade the immune system and spread through tissues. Released exotoxin causes necrosis of the surrounding tissue. The boxcar-shaped bacteria themselves produce gas, which leads to a bubbly deformation of infected tissues. Treatment of histotoxic *C. perfringens* consists of penicillin G (to kill the organism), hyperbaric O_2, administration of antitoxin, and debridement of infected areas.

C. difficile is a motile, obligate anaerobic or microaerophilic, gram-positive, spore-forming, rod-shaped bacillus. *C. difficile*–associated disease occurs when the normal intestinal flora is altered, allowing the bacteria to flourish in the intestinal tract and produce a toxin that causes a watery diarrhea. *C. difficile* is recognized as a chief cause of nosocomial (hospital-acquired) diarrhea. Infections can appear through the use of broad-spectrum antibiotics, which lower the relative amount of other normal gut flora, thereby allowing *C. difficile* proliferation and infection into the large intestine. The bacterium then releases two enterotoxins (colitis toxins A and B) that are cytotoxic to enterocytes, thus causing a pseudomembranous colitis by destroying the intestinal lining.

The end result is diarrhea. The preferred method of treatment is oral vancomycin or metronidazole plus rehydration.

Actinomyces

Actinomyces spp. are gram-positive, obligate anaerobes known to reside in the mouth and intestinal tract. They are morphologically similar to fungus in that they form filamentous branches. Pathology due to proliferation of organisms usually occurs following injury or trauma to tissue, resulting in actinomycosis (abscess formation and swelling at the site of infection). Microscopic examination of pus reveals exudates with granular texture caused by sulfur granules, resulting from the bacterium and its waste. *A. israelii* is most commonly associated with actinomycosis; however, other *Actinomyces* bacteria are capable of causing disease. Actinomycosis can be treated with penicillin.

Bifidobacterium

Bifidobacteria are anaerobic, gram-positive bacilli rarely associated with infection. *Bifidobacterium dentium*, a normal inhabitant of the gut flora, is the only pathogenic species reported. Microscopically, these organisms appear bone shaped, making them easy to identify. They are obligate anaerobes and require very low O_2 tension to survive and achieve moderate growth.

Eubacterium and *Propionibacterium*

Eubacterium spp. are of only minor clinical importance. They are normal flora of the intestinal tract and cause infection under opportunistic conditions. *Eubacterium lentum* is the species that is most often isolated; it has been linked to endocarditis. Biochemical testing can distinguish *Eubacterium* from the other gram-positive, anaerobic rods. *Propionibacterium* spp. are common gram-positive anaerobes isolated in the laboratory. *Propionibacterium acnes* is typically noninvasive and harmless, but it has pathogenic potential and has been linked to endocarditis, wound infections, and abscesses. Despite its name, *P. acnes* is not the causative agent of acne, although it may infect acne sites. Microscopically, *Propionibacterium* clumps together and shows a minor branching tendency, with uneven gram-staining patterns. Colonies grow best in anaerobic or microaerophilic environments on blood agar.

●●● AEROBIC GRAM-NEGATIVE BACILLI

Enterobacteriaceae

Members of the Enterobacteriaceae family are among the most pathogenic and commonly encountered organisms in clinical microbiology. They are large, gram-negative rods usually associated with intestinal infections and also cause meningitis, bacillary dysentery, typhoid, and food poisoning. They are oxidase negative, and all members of this family are glucose fermenters and nitrate reducers. The pathogenicity of each enteric member may be determined by its ability to metabolize lactose. The various genera of the Enterobacteriaceae family most commonly encountered in the clinical laboratory are presented here.

> **HISTOLOGY**
>
> **Histology of the Intestine**
>
> The intestine is histologically characterized by the presence of the muscularis propria with the Auerbach plexus between the inner circular and outer longitudinal smooth muscle layers. The Meissner plexus is similar in function but found in submucosal areas.

Escherichia coli

Escherichia coli is the main cause of human urinary tract infections, and it has been linked to sepsis, pneumonia, meningitis, and traveler's diarrhea. It is part of the normal flora of the intestinal tract. *E. coli* produces vitamin K in the large intestine, which plays a crucial role in food digestion. Pathogenic strains of *E. coli* have a powerful cell wall–associated endotoxin that causes septic shock, and two enterotoxins. One enterotoxin is a heat-labile (LT) molecule that stimulates adenosine diphosphate ribosylation via adenylate cyclase activity, leading to dysregulation of chloride ions in the gut. A heat-stable toxin (ST) also contributes to diarrheal illness. Treatment of *E. coli* infections with antibiotics sometimes leads to release of additional factors causing severe shock, which is potentially fatal. At the species level, *E. coli* and *Shigella* are indistinguishable, with much overlap between diseases caused by the two organisms.

Four etiologically distinct diseases are defined according to clinical symptoms. Enteropathogenic *E. coli* is commonly found associated with infant diarrhea due to destruction of microvilli without invasion of the organism, leading to fever, diarrhea, vomiting, and nausea, usually with nonbloody stools. Enterotoxigenic *E. coli* is the cause of traveler's diarrhea due to the plasmid-encoded LT and ST toxins. Enteroinvasive *E. coli* produces a dysentery indistinguishable clinically from shigellosis. Enterohemorrhagic *E. coli*, usually of serotype O157:H7, produces a hemorrhagic colitis characterized by bloody and copious diarrhea with few leukocytes in afebrile patients.

Shigella

Shigella spp. are closely related to *Escherichia*. *Shigella* is usually distinguishable from *E. coli* by virtue of the fact that it is anaerogenic (does not produce gas from carbohydrates) and lactose negative. *Shigella* is an invasive, facultative, gram-negative rod pathogen; four species may be designated based on serologic identity, all of which cause bloody diarrhea accompanied by fever and intestinal pain. The members of the species causing shigellosis are *Shigella dysenteriae* (serotype A), *Shigella flexneri* (serotype B), *Shigella boydii* (serotype C), and *Shigella sonnei* (serotype D). Serotype D is primarily responsible for shigellosis. Following infection, dysentery results from bacterial damage of epithelial layers lining the intestine, with release of mucus and blood and attraction of leukocytes. A neurotoxic, enterotoxic, cytotoxic, chromosome-encoded shiga toxin is responsible for the pathology. The toxin inhibits protein synthesis. Managing dehydration is of primary concern. Indeed, mild diarrhea often is not recognized as shigellosis. Patients with severe dysentery are usually treated with antibiotics (e.g., ampicillin).

Salmonella

Salmonella spp. are facultative, gram-negative, non–lactose-fermenting rods. Transmission of *Salmonella* occurs through ingestion of uncooked meats and eggs; chickens serve as a major reservoir in the food chain. Ingestion of contaminated foods can cause intestinal infection leading to diarrhea, vomiting, and chills. Pathogenic entry occurs with the help of M cells, which are able to translocate organisms across enteric mucosa. *Salmonella* spp. are classified according to their surface antigens. In the United States, *S. typhimurium* (gastroenteritis) and *S. enteritidis* (enterocolitis) are the two leading causes of salmonellosis (inflammation of the intestine caused by *Salmonella*). *S. typhi* causes typhoid fever (enteric fever), which is characterized by fever, diarrhea, and inflammation of infected organs. Most *Salmonella* infections can be treated with ciprofloxacin or ceftriaxone.

Yersinia

Yersinia is an invasive pathogen that can infiltrate the intestinal lining to enter the lymphatic system and the blood supply. Infection through ingestion of contaminated foods causes severe intestinal inflammation (yersiniosis). *Y. enterocolitica* is a urease-positive organism associated with diarrhea, fever, and abdominal pain (gastroenteritis) caused by release of its enterotoxin. A similar but less severe disease is caused by *Y. pseudotuberculosis*. *Y. pseudotuberculosis* (formerly called *Pasteurella pseudotuberculosis*) is pathogenic, causing mesenteric lymphadenitis in humans. Antibiotic treatment consists of aminoglycosides, chloramphenicol, or tetracycline.

Although not a true enteric pathogen, *Y. pestis* has historical significance as the causative agent of bubonic, pneumonic, and septicemic plagues. Human contraction of **bubonic plague** may occur via flea bites, with transfer of disease from a rodent reservoir. *Y. pestis* is a urease-negative organism that has cell wall protein-lipoprotein complexes (V and W antigens) that inhibit phagocytosis, and it releases a toxin during infection that inhibits electron transport chain function. Swelling of the lymph nodes and delirium are observed within a few days of infection, followed by pneumonia characterized by high fever, cough with bloody sputum, chills, and severe chest pains. Death will occur if it is left untreated. Effective antibiotic treatment consists of streptomycin and gentamicin.

Other Pathogenic Enterobacteriaceae

Edwardsiella spp. are biochemically similar to *E. coli*; however, *Edwardsiella tarda* has the distinction of producing hydrogen sulfide; it can cause gastroenteritis and infect open wounds in humans. *Citrobacter* is part of the normal gut flora; *Citrobacter freundii* can cause diarrhea and possibly extraintestinal infections. *C. diversus* may cause meningitis in newborns. *Klebsiella* spp. are large, nonmotile bacteria that produce a heat-stable enterotoxin. *Klebsiella pneumoniae* causes pneumonia with characteristic bloody sputum, and urinary tract infections in catheterized patients. *Enterobacter* includes multiple species of highly motile bacteria that normally reside in the intestinal tract. They are biochemically similar to *Klebsiella* and can cause opportunistic infections of the urinary tract. *Enterobacter aerogenes* and *E. cloacae* are two examples of pathogens that are associated with urinary tract and respiratory tract infections. Members of the *Serratia* genus produce pathogenic enzymes including DNase, lipase, and gelatinase. *Serratia marcescens* causes urinary tract infections, wound infections, and pneumonia. *Proteus* spp. are highly motile and form irregular "swarming" colonies. *Proteus mirabilis* and *P. vulgaris* cause wound and urinary tract infections, especially important in the immunocompromised or immunosuppressed host. Of the *Morganella* spp., *Morganella morganii* is clinically important and can cause urinary tract and wound infections as well as diarrhea. Finally, *Providencia* spp. have been associated with nosocomial (hospital-acquired) urinary tract infections; *Providencia alcalifaciens* has been associated with diarrhea in children.

Haemophilus

Respiratory tract infections caused by pleomorphic, aerobic, gram-negative rods include organisms of the *Haemophilus*, *Legionella*, and *Bordetella* spp. The *Haemophilus* genus represents a group of gram-negative rods that grow on blood agar, requiring blood factors X (an iron tetrapyrrole such as hemin) and V (oxidized nicotinamide adenine dinucleotide or reduced nicotinamide adenine dinucleotide phosphate). Morphologically, *Haemophilus* bacteria usually appear as tiny coccobacilli, designated as pleomorphic bacteria because of their multiple morphologies. *Haemophilus* spp. are classified by their capsule into six different serologic groups (a to f). Infection by *H. influenzae* is common in children and causes secondary respiratory infections in individuals who already have the flu. *H. influenzae* may present with or without a pathogenic polysaccharide capsule and is present as normal flora residing in the nose and pharynx. Strains without a capsule usually cause mild, contained infections (otitis media, sinusitis); however, type b encapsulated *H. influenzae* can cause meningitis with fever, headache, and stiff neck. Other presentations occur as cellulitis, arthritis, or sepsis. Before the introduction of a highly effective vaccine, *H. influenzae* was the most common cause of bacterial meningitis in children younger than 5 years in developed countries (*S. pneumoniae* and *N. meningitides* now are more important). In less well-developed countries, *H. influenzae* infection is still a major problem. Respiratory infection may spread from the blood to eventually infect CNS tissue. *Haemophilus* infection is typically associated with other lung disorders (chronic bronchitis, pneumonia) as well as with bacteremia and conjunctivitis. Cephalosporins are used in treatment. Other species of clinical interest are *H. aegyptius*, which cause pinkeye (conjunctivitis), and *H. ducreyi*, which causes a sexually transmitted disease characterized by painful genital ulcers (chancroid).

Legionella

The genus *Legionella* was headlined in the mid-1970s when an outbreak of pneumonia at an American Legion convention led to multiple deaths (Legionnaires' disease). The causative agent, *Legionella pneumophila*, is a gram-negative intracellular bacterium that produces off-white, circular colonies. Respiratory transmission leads to infection characterized by the gradual onset of fever, chills, and a dry cough; eventual progression to severe pneumonia may occur, with possible spread to the gastrointestinal tract and CNS. Advanced infections are characterized by diarrhea, nausea, disorientation, and confusion. *L. pneumophila* is associated with Pontiac fever, evidenced by generally mild flulike symptoms that do not develop or spread beyond the lungs. *L. pneumophila* infections are easily treated with erythromycin. *L. micdadei* is similar to *L. pneumophila* but does not produce β-lactamase. *L. micdadei* causes similar flulike symptoms and pneumonia.

Bordetella

Bordetella organisms are small, gram-negative coccobacilli. They are strict aerobes. The most clinically important species is *Bordetella pertussis*, which causes whooping cough. The organism enters the respiratory tract after inhalation and destroys the ciliated epithelial cells of the trachea and bronchi through various toxins. These toxins include the pertussis toxin (exotoxin) that activates host cell production of cAMP to modulate cell protein synthesis regulation, a tracheal cytotoxin that causes destruction of ciliated epithelial cells, and a cell surface hemagglutinin to assist in bacterial binding to the host cells. Antimicrobial therapy for whooping cough usually consists of erythromycin.

Two other species of *Bordetella* of clinical importance are *B. parapertussis*, a respiratory pathogen that causes mild pharyngitis, and *B. bronchiseptica*, which causes pneumonia and otitis media.

●●● ORGANISMS OF ZOONOTIC ORIGIN

Pasteurella

Infection of the lungs with *Pasteurella* spp., usually *Pasteurella multocida* or *P. haemolytica*, causes pneumonic pasteurellosis, a fulminating, fatal lobar pneumonia. Other pathologies attributed to these organisms include septicemic pasteurellosis and a hemorrhagic septicemia. *P. multocida* is a member of the genus of gram-negative, facultatively anaerobic, ovoid to rod-shaped bacteria of the family Pasteurellaceae. It is an extracellular parasite that may be cultured on chocolate agar and typically produces a foul odor. *P. multocida* commonly infects humans and is acquired usually through scratches or bites from cat or dogs. Patients tend to exhibit swelling, cellulitis, and some bloody drainage at the wound site, as well as abscesses and septicemias. Infection in nearby joints can cause swelling and arthritis. *P. haemolytica*, a species that is part of the normal flora of cattle and sheep, is the etiologic agent of hemorrhagic septicemia. Both *P. multocida* and *P. haemolytica* are susceptible to penicillin, tetracycline, and chloramphenicol.

Brucella

Brucella is an aerobic, gram-negative coccobacillus that is the causative agent of brucellosis. Four species normally found in animals can infect humans: *Brucella abortus* (cattle), *B. suis* (swine), *B. melitensis* (goats), and *B. canis* (dogs). *Brucella* enters the body by way of the skin, digestive tract, or respiratory tract, after which it may enter the blood and lymphatics. It is an intracellular pathogen that multiplies inside phagocytes to eventually cause bacteremia (bacterial blood infiltration). Symptoms include fever, sweats, malaise, anorexia, headache, myalgia, and back pain. In the undulant form (less than 1 year from illness onset), symptoms include fevers, and arthritis, with possible neurologic manifestation in a small number of cases. In the chronic form, symptoms can include chronic fatigue syndrome with accompanying depression and eventual arthritis. Afflicted individuals are successfully treated with streptomycin or erythromycin.

Francisella

Francisella tularensis is a small, gram-negative, aerobic bacillus. The two main serotypes are Jellison types A and B. Type A is the more virulent form; infection through tick bite or direct contact will lead to tularemia. *F. tularensis*, also referred to as *Pasteurella tularensis*, causes sudden fever, chills, headaches, diarrhea, muscle aches, joint pain, dry cough, and progressive weakness. The disease also can be contracted by ingestion or inhalation. Tularemia occurs in six different forms: typhoidal, pneumonic, oculoglandular, oropharyngeal, ulceroglandular, and glandular. Treatment includes a regimen of streptomycin or gentamycin.

Bartonella

Bartonella henselae is a fastidious, gram-negative bacterium that is the cause of many diseases such as bacillary angiomatosis, visceral peliosis, septicemia, endocarditis, and cat-scratch disease. The most common symptoms are persistent fever lasting up to 8 weeks, abdominal pain, and lesions around sites of infection. Aminoglycosides and rifampin are effective and bactericidal, whereas β-lactams are ineffective in treatment.

Vibrio

The *Vibrio* genus contains motile, gram-negative bacteria that are obligate aerobes. They are comma-shaped rods with a single polar flagellum, facultative anaerobes that are oxidase positive. Although *Vibrio* spp. are noninvasive pathogens, they cause severe diarrheal illness and thousands of deaths annually. The organisms are waterborne and are transmitted to humans through ingestion of infected water or through fecal transmission.

Vibrio cholerae is the causative agent of cholera, characterized by severe diarrhea with a rice-water color and consistency. Sixty percent of cholera deaths are due to dehydration. Ingested organisms descend to the intestinal tract, bind to epithelium, and subsequently release an exotoxin (choleragen) (Fig. 12-5), causing water to passively flow out of cells. It is critical to replace fluids and electrolytes when treating cholera patients. *V. cholerae* is susceptible to administration of doxycycline or tetracycline, as are other members of this species. *V. parahaemolyticus* is another species that causes diarrhea as well as cramps, nausea, and fever. The disease is transmitted by eating infected seafood and is self-limiting to about 3 days. *V. vulnificus* and *V. parahaemolyticus* may also be contracted from contaminated seafood. Unlike other *Vibrio* spp., *V. vulnificus* is invasive and able to enter the bloodstream through the epithelium of the gut. Fever, vomiting, and chills are the symptoms; wound infections can occur with resulting cellulitis or ulcer formations.

●●● CAMPYLOBACTER AND HELICOBACTER

Two groups of gram-negative organisms, *Campylobacter* and *Helicobacter*, may be found residing in gut tissue. Both are curved or spiral shaped as well as motile and catalase positive; they are genetically related. Organisms of the genus *Campylobacter* are gram-negative microaerophiles that cause diarrhea. They achieve cell motility by way of polar flagella. *Campylobacter jejuni* causes gastroenteritis and is usually acquired by eating undercooked food or drinking contaminated milk or water. Symptoms of infection are fever, cramps, and bloody diarrhea, which is caused by penetration of the lining of the small intestine. It can be treated with antibiotics (erythromycin) but is usually self-limited.

Helicobacter spp. also are gram-negative, microaerophilic organisms. *Helicobacter pylori* is a spiral-shaped bacterium that is found in the gastric mucus layer or adherent to the epithelial lining of the stomach. *H. pylori* causes more than 90% of duodenal ulcers and up to 80% of gastric ulcers. The mechanisms of pathogenesis for ulceration remain incompletely defined. The organism characteristically produces a urease that generates ammonia and CO_2. Infected patients can be treated with an antacid as well as tetracycline to treat the ulcers and inhibit the growth of the organism.

●●● NONFERMENTERS

The nonfermenters are gram-negative rods that either do not ferment glucose for energy or do not use glucose at all. *Pseudomonas* and *Acinetobacter* spp. fall into this category. Pseudomonads are motile organisms that use glucose oxidatively. *Pseudomonas* comprises five groups based on ribosomal

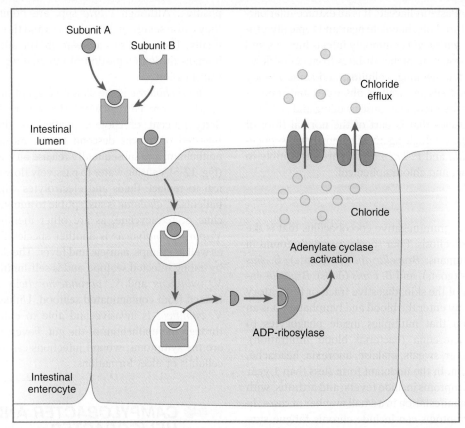

Figure 12-5. *V. cholerae* organisms adhere to the intestinal microvilli upon which cholera toxin is secreted. The toxin contains two subunits (A and B); the B subunit binds to gangliosides on epithelial cell surfaces, allowing internalization of the A subunit. B subunits provide a hydrophobic channel through which A penetrates, following which the A subunit catalyses ADP-ribosylation to activate adenylate cyclase in cell membranes of gut epithelium. The end result is massive secretion of ions and water into the gut lumen with resultant dehydration.

RNA (rRNA)/DNA homology. These bacteria are often encountered in hospital settings and are a major source of nosocomial infections. Clinically, they are resistant to most antibiotics and are capable of surviving harsh conditions. Both organisms target immunocompromised individuals, burn victims, and individuals on respirators or with indwelling catheters. Pseudomonads produce an alginate slime layer that is resistant to phagocytosis. They readily colonize the lungs of cystic fibrosis patients, increasing the mortality rate of these individuals. Infection may occur in multiple tissues, leading to urinary tract infections, sepsis, pneumonia, and pharyngitis.

Pseudomonas

Pseudomonas aeruginosa is commonly isolated from clinical specimens (wound, burn, and urinary tract infections). It may be found widely distributed in soil and water. Its pathogenicity involves bacterial attachment and colonization, followed by local invasion, and dissemination with systemic disease. A surface lipopolysaccharide layer assists in adherence to host tissues and prevents leukocytes from ingesting and lysing the organism. Lipases and exotoxins then proceed to destroy host cell tissue. In healthy children, diseases are limited primarily to those associated with attachment and local infection (e.g., otitis externa, urinary tract infections, dermatitis, cellulitis). In immunocompromised hosts and neonates who do not have a fully competent immune response, infection may appear as a disseminated infection (e.g., pneumonia, endocarditis, peritonitis, meningitis, overwhelming septicemia).

P. aeruginosa and *P. maltophilia* account for 80% of opportunistic infections by pseudomonads. *Burkholderia* (*Pseudomonas*) *cepacia* is a related opportunistic pathogen of cystic fibrosis patients that can be distinguished from *Pseudomonas* spp. because it is lysine positive. The spread of *Pseudomonas* is best controlled by cleaning and disinfecting medical equipment. Pseudomonads typically are resistant to multiple therapeutic and antimicrobial agents; therefore antibiotic susceptibility testing of clinical isolates is mandatory. In general, the combination of gentamicin and carbenicillin can be effective in treatment of acute infections.

Acinetobacter

Acinetobacter spp. are oxidase-negative, nonmotile bacteria. They appear as gram-negative coccobacilli when viewed microscopically. *Acinetobacter* poses little risk to healthy people; individuals with weakened immune systems, chronic lung disease, or diabetes may be more susceptible to infection. *Acinetobacter baumannii* accounts for about 80% of reported infections and is linked to hospital-acquired infections of the skin and open wounds. In addition, *A. baumannii* is a major agent leading to pneumonia and meningitis. *A. lwoffii* is the causative agent for the majority of reported meningitis caused by *Acinetobacter*. Owing to high resistance to multiple antibiotics, the combination of an aminoglycoside and a second agent is usually recommended for treatment.

Although not of the same genus, *Stenotrophomonas maltophilia* (formerly known as *Xanthomonas maltophilia*) is similar to the pseudomonads. This motile bacterium is a cause of nosocomial infections in immunocompromised patients. *Flavobacterium* spp. are ubiquitous organisms that can cause infection in premature infants and immunocompromised individuals. Most species metabolize glucose oxidatively; all species are motile and oxidase positive. Of clinical importance, *Flavobacterium meningosepticum* causes neonatal meningitis and is typically penicillin resistant.

●●● ANAEROBIC GRAM-NEGATIVE BACILLI

Bacteroides

Bacteroides spp. are anaerobic bacteria that inhabit the digestive tract; interestingly, 50% of fecal matter is composed of *Bacteroides fragilis* cells! *Bacteroides* organisms are the anaerobic counterpart of *E. coli*. They grow well on blood agar. Microscopically, they appear to contain large vacuoles similar in appearance to spores. *Bacteroides* spp. produce a very large capsule but are not spore forming. They do not possess an endotoxin in their cell membrane, which limits their pathogenicity. Infection occurs after severe trauma to the gut and abdominal region, resulting in abscess formation with accompanying fever. Antibiotic treatment consists of metronidazole or clindamycin.

Fusobacterium

Fusobacterium organisms are anaerobic, gram-negative bacilli that are similar to certain *Bacteroides* spp. They appear as spindle-shaped cells with sharp ends. Both reside in the gut but are capable of causing serious infection. *Fusobacterium* spp., the most common of which is *Fusobacterium nucleatum*, are associated with pleuropulmonary infections and congestion exhibited as sinusitis. They are also capable of causing infection in the oral cavity (the mouth) and may be a major cause of gingivitis.

●●● SPIROCHETES

Spirochetes are long, slender bacteria, usually only a fraction of a micron in diameter but anywhere from 5 to 250 µm long. The best known spirochetes of clinical importance are those that cause disease. Among spirochetal diseases are syphilis and Lyme disease.

Treponema

Treponema pallidum is the causative organism of syphilis. It is a motile spirochete that is generally acquired by close sexual contact and which enters host tissue by breaches in squamous or columnar epithelium. Disease is marked by a primary chancre (an area of ulceration and inflammation) seen in genital areas, which manifests soon after the primary infection. Progression to secondary and tertiary syphilis is marked by maculopapular rashes and eventual granulomatous response with CNS involvement. Nonvenereal treponemal diseases include pinta, caused by *Treponema carateum*, and disfiguring yaws, caused by *Treponema pallidum* ssp. *pertenue*.

Borrelia

Borrelia burgdorferi is the spirochete that causes Lyme disease. In contrast to *T. pallidum*, *Borrelia* has a unique nucleus containing a linear chromosome and linear plasmids. *Borrelia* is transmitted by tick bites (*Ixodes*) during blood feeding. An early indication of Lyme disease is a distinctive skin lesion called *erythema migrans*. If it is left untreated, an erosive arthritis similar to rheumatoid arthritis can occur and, eventually, chronic progressive encephalitis and encephalomyelitis. A number of other *Borrelia* spp. can cause endemic relapsing fever, with causative agents including *B. duttonii*, *B. hermsii*, and *B. dugesi*.

Leptospira

Leptospira spp. (leptospires) are long, thin motile spirochetes. They are the causative agent of leptospirosis, a febrile illness that may lead to aseptic meningitis if left untreated. Symptoms of infection include fever, chills, and headache, with occasional presentation of jaundice. Organisms can be spread in water contaminated by infected animal urine.

> **PATHOLOGY**
>
> **Erythema Migrans of Lyme Disease**
>
> Erythema migrans (also called *erythema chronicum migrans*) is the skin lesion that develops at the site of a bite from a deer tick infected with borreliosis. The early lesion is characterized by an expanding area of red rash, often with a pale center ("bulls-eye") at the site of the tick bite, indicative of early signs of Lyme disease.

●●● OBLIGATE INTRACELLULAR PATHOGENS

Chlamydia

Obligate intracellular pathogens of clinical importance include the *Chlamydia* and *Rickettsia* spp. *Chlamydia* are small, obligate, intracellular parasites that were once considered to be viruses. The family Chlamydiaceae consists of the one genus *Chlamydia* with three species that cause human disease. *Chlamydia trachomatis* can cause urogenital infections, trachoma, conjunctivitis, pneumonia, and lymphogranuloma venereum (LGV). *C. pneumoniae* can cause bronchitis, sinusitis, and pneumonia. *C. psittaci* can cause pneumonia (psittacosis). *Chlamydia* spp. have an inner and outer membrane and a glycolipid but not a peptidoglycan layer. They are unable to make their own adenosine triphosphate and thus are energy parasites. The structure of all three species is similar. The infectious agent is the elementary body form, which is characterized by a rigid outer membrane that is extensively cross-linked by disulfide bonds. The elementary bodies bind to receptors on host epithelial cells to initiate infection (Fig. 12-6). Metabolically active, replicating intracellular forms are referred to as reticulate bodies. Reticulate bodies possess a fragile membrane lacking the extensive disulfide bonds characteristic of the elementary body. Human infectious biovars have been subdivided into serovars (serologic variants) that differ in major outer membrane proteins. The *C. trachomatis* serovars A, B, and C are associated with ocular disease; serovars D through K are associated with conjunctivitis, urethritis, cervicitis, and pneumonia; and serovars L1, L2, and L3 are associated with lymphogranuloma venereum. *C. pneumoniae* is the causative agent of an atypical pneumonia similar to that caused by *Mycoplasma pneumoniae*. The organism is transmitted person-to-person by respiratory droplets. For all the *Chlamydia*, effective treatment includes tetracyclines, erythromycin, or sulfonamides.

Rickettsia, *Ehrlichia*, and *Coxiella*

The genera *Rickettsia*, *Ehrlichia*, and *Coxiella* are a diverse collection of gram-negative, obligate, intracellular bacteria found in arthropod vectors (ixodid ticks, lice, and fleas). They are considered zoonotic pathogens. They infect white blood cells, causing blood-borne, disseminated infections of endothelium and vascular smooth muscle. *Rickettsia rickettsii* causes Rocky Mountain spotted fever and a form of rickettsial pox. *Ehrlichia* spp. cause ehrlichioses. *Coxiella* spp. (*Coxiella burnetii*) cause Q fever, which manifests as an acute febrile illness with pneumonia or as a chronic infection with accompanying endocarditis. All these species are sensitive to doxycycline and tetracycline.

●●● ACID-FAST ORGANISMS

A number of clinically relevant organisms are described as *acid-fast* owing to their staining properties with carbolfuchsin stain (Ziehl-Neelsen stain); only acid-fast organisms retain the stain after treatment with an acid alcohol. A brilliant red coloration results from retention of carbolfuchsin within the cell membrane. Although the exact molecular mechanism

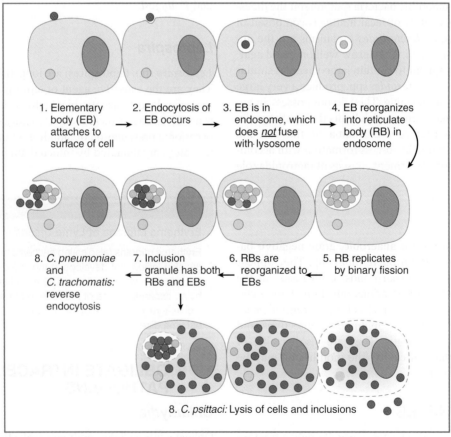

Figure 12-6. Chlamydial developmental cycle. Infectious elementary bodies (*EB*) bind to receptors on susceptible epithelial cells and are internalized. Inside the host cell endosomes, reorganization into the reticulate body (*RB*) form occurs. This is accompanied by inhibition of the fusion of endosomes with lysosomes to limit intracellular killing. RBs replicate by binary fission and finally reorganize into rigid EBs; inclusions containing up to 500 progeny are extruded by reverse endocytosis or by cell death and subsequent cell lysis.

for dye retention is unknown, this group of organisms is characterized by a high content of mycolic acids within cellular membranes. Of these organisms, the mycobacteria are of major clinical relevance with nearly one third of the world's population infected. *Nocardia* spp. and intestinal coccidia also belong to this group; pathologic manifestations of infection caused by these organisms are discussed.

Mycobacterium

As discussed in Chapter 10, organisms that are the causative agents of tuberculosis and leprosy have evolved to inhibit normal macrophage killing mechanisms (e.g., phagosome-lysosome fusion) and survive within the "disarmed" professional phagocyte. This leads to development of hypersensitive pathologies, established to contain a central nidus of infection. Mycobacteria are nonmotile, slow-growing, rod-shaped organisms that are obligate aerobes. They have a cell envelope with a high lipid content and contain complex, long-chain fatty acids (mycolic acids) that are otherwise found only in *Nocardia* and *Corynebacterium*. Mycobacteria are catalase positive, with the exception of nonpathogenic *M. kansasii* and isoniazid-resistant *M. tuberculosis*.

Mycobacterium tuberculosis

M. tuberculosis (MTB) is a major health concern, with an estimated 8.4 million new cases a year and 2 to 3 million deaths worldwide. The organism is acquired through inhalation of aerosolized infected droplets. It has an extremely slow growth rate (doubling time of 18 hours), and a cell envelope that is rich in waxes and lipids, especially mycolic acids that are covalently linked to arabinogalactans (Fig. 12-7). Once inside the host, the organism is engulfed by macrophages, stimulating responses leading to exudative (pneumonia-like) or granulomatous lesions. The granulomatous lesions are characterized by giant multinucleated cells, surrounded by lymphocytes, forming the basis of a tubercle. Necrotic events occur as the organisms persist, leaving a caseous center to the granulomatous response. Eventual erosion of lesions into bronchial airways leads to spread of disease. Resolution may occur, leaving remnant fibrotic and calcified lesions, which are referred to as *Ghon complexes*. Most individuals successfully contain organisms. However, immunocompromised individuals (due to HIV infection, malnutrition, old age, or iatrogenic immunosuppression) can undergo reactivation events, with infection dissemination and reseeding of organisms to apical lung tissue. The pathology of mycobacterial infections is quite complex, with postprimary tuberculosis progressing to produce caseous pneumonia and cavitary disease. The clinical outcome of infection is due to the nature of the host response.

Exposure to mycobacterial surface antigens leads to responses that can be detected using the tuberculin skin test, functionally detecting an inducible delayed-type hypersensitive response (see Chapter 7). Small amounts of purified protein derivative (Mantoux test) are injected intradermally; induration occurring within 24 hours signifies positive exposure to organisms. Because of the slow growth of the organisms and the presence of the waxy cell envelope, therapeutics to combat infection must

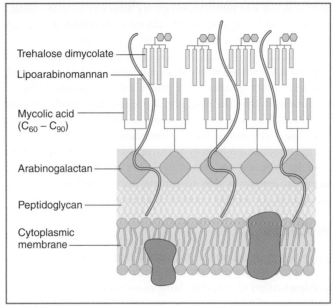

Figure 12-7. The mycobacterial cell envelope is rich in waxes and lipids, especially mycolic acids, which are covalently linked to arabinogalactans. The unique long-chain mycolic acid trehalose 6,6'-dimycolate (C_{60}–C_{90}) is responsible for "cording" seen with virulent strains, giving colonies a serpentine morphology. Mycolic acids form an asymmetric lipid bilayer with shorter-chain glycerophospholipids. The cell envelope also contains peptidoglycan and has a complex glycolipid structure forming the outermost layer, which sits atop the peptidoglycan and lipid bilayer of the cytoplasmic membrane. Other important components include lipoarabinomannan and arabinogalactan.

be given for long periods. It is typical to administer 6 months of therapy with a first-line antituberculosis agent, such as isoniazid, rifampin, pyrazinamide, streptomycin, or ethambutol. In cases of drug resistance, a second-line antituberculosis drug may also be prescribed, such as para-aminosalicylic acid, rifabutin, ethionamide, kanamycin, or a fluoroquinolone.

Atypical Mycobacteria

The atypical mycobacteria include *M. kansasii*, which causes a lung disease clinically resembling MTB, especially in individuals with preexisting lung conditions. *M. marinum* causes "swimming pool granulomas" and abscesses. *M. avium-intracellulare* complex (MAC) is pleomorphic, with multiple disease-causing serovars included in this group. MAC infections typically cause pulmonary disease similar to MTB in the immunocompromised host; disseminated infection is typical in individuals with complicating HIV infections. *M. fortuitum* and *M. chelonei* are saprophytes that rarely cause disease; however, infections have been reported in individuals with joint replacements.

Mycobacterium leprae

M. leprae causes leprosy, also known as *Hansen disease*. The optimal temperature for growth is lower than core body temperature, so it grows on skin and superficial nerves, infecting macrophages and Schwann cells. *M. leprae* has resisted efforts for growth in culture and thus is difficult to study. There are two clinical spectra of disease—tuberculoid and lepromatous leprosy—based on immune function and host response. The high resistance of tuberculoid leprosy is associated with cellular response and formation of prominent granulomatous lesions. Delayed hypersensitivity skin tests are intact, and there is predominant hyperplasia of the lymph nodes. Lepromatous leprosy is associated with the accumulation of highly activated, "foamy" macrophages filled with viable organisms. Delayed hypersensitivity skin tests are depressed; the levels of antibodies are high and vascular lesions occur (erythema nodosum leprosum). Borderline leprosy has intermediate findings. Clinically, lepromatous leprosy is characterized by symmetric skin nodules, plaques, and loss of eyelashes and body hair. Loss of digits in leprosy is due to trauma and secondary infection. Treatment for leprosy consists of dapsone and clofazimine.

Nocardia

Respiratory infection due to *Nocardia* may occur through inhalation, whereas primary cutaneous disease results from soil contamination of wounds. Most cases present as an invasive pulmonary disorder with the potential for disseminated disease that may lead to brain abscess; however, 20% of cases present as cellulitis. *Nocardia* appear similar to fungal agents, with beaded, branching filaments. *Nocardia asteroides* is the major organism accounting for infections, although other pathogenic species have been identified, including *N. farcinica*, *N. nova*, *N. transvalensis*, *N. brasiliensis*, and *N. pseudobrasiliensis*.

Acid-Fast Intestinal Coccidia

Three intestinal coccidia infecting humans stain acid-fast and should be considered unicellular protozoa. They are *Cryptosporidium parvum*, *Cyclospora cayetanensis*, and *Isospora belli*. *C. parvum* is the etiologic agent of diarrheal disease (cryptosporidiosis) caused by members of the genus *Cryptosporidium*. The most common symptom of cryptosporidiosis is watery diarrhea with accompanying dehydration, weight loss, stomach cramps or pain, fever, nausea, and vomiting. *Cryptosporidium* is resistant to chlorine and can be spread in pool water. *Cryptosporidium cayetanensis* is the agent causing cyclosporiasis. *Cyclospora* infects the small intestine (bowel), resulting in watery diarrhea with frequent, sometimes explosive, bowel movements. The agent for isosporiasis, *Isospora belli*, infects epithelial cells of the small intestine and is the least common of the three intestinal coccidia.

●●● MYCOPLASMA

Mycoplasma pneumoniae is a small (0.3-μm diameter), wall-less organism that is transmitted through aerosolized droplets. *Mycoplasma* binds firmly to respiratory epithelium; it is the causative agent of nonviral atypical pneumonia symptomatically characterized by sore throat, headache, myalgia, and whitish, nonbloody sputum. Infection with *M. pneumoniae* can lead to autoantibody production (usually IgM isotypes), producing cross-reactive antibodies directed against red cell antigens. The autoantibodies are referred to as cold agglutinins and cause a reactive, autoimmune, hemolytic disease. Because *Mycoplasma* has no cell wall, agents such as penicillin and cephalosporins are not effective.

KEY POINTS

- Clinical manifestation due to bacterial infection depends in many ways on virulence factors produced by the bacteria that mediate environmental conditions and affect host immune function.
- The nonmotile non–spore-forming gram-positive organisms are a heterogenous collection of agents that colonize humans. Many cause pyogenic (producing pus) and pyrogenic (producing fever) infections. Examples include *Staphylococcus*, *Streptococcus*, and *Enterococcus* spp., which represent a major population responsible for cutaneous infections and systemic disease.
- The gram-negative cocci, such as *Neisseria*, include aerobic pathogens commonly found on mucosal membranes with the ability to cause disease in both healthy and immunocompromised individuals.
- Aerobic gram-positive bacilli, including *Bacillus*, *Lactobacillus*, *Listeria*, and *Corynebacteria* spp., are subdivided according to shape, virulence, and epidemiology. The anaerobic, gram-positive bacilli include agents such as the *Clostridium*, *Actinomyces*, and *Propionibacterium* spp. In some cases, such as *Clostridium*, production of varied enterotoxins and neurotoxins contributes to severe pathogenesis upon infection.
- Aerobic gram-negative bacilli include *E. coli* and the closely related *Shigella*. *E. coli* is the main cause of human urinary tract infections, and it has been linked to sepsis, pneumonia, meningitis, and traveler's diarrhea. *Salmonella* spp., representing non–lactose-fermenting rods, cause of host of disorders, most of which are characterized by fever, diarrhea, and inflammation. Other organisms in this category include the Enterobacteriaceae, which include *Haemophilus*, *Legionella*, and *Bordetella*.
- Organisms of zoonotic origin represent those that cross species, generally through contaminated waste materials or via other contact with infected animals. *Pasteurella*, *Brucella*, and *Bartonella* are examples of these infectious species.
- *Helicobacter* and *Campylobacter* are gram-negative species that infect gut mucosa.
- Nonfermenters include certain gram-negative rods that exemplify gram-negative, motile, non–spore-forming, rod-shaped bacteria that cause a variety of infectious diseases. *Pseudomonas* and *Acinetobacter* are in this category.
- The spirochetes are slender, spiral, motile bacteria. Organisms such as *Treponema* can cause syphilis, relapsing fever, and yaws. Another member is *Borrelia*, the agent responsible for Lyme disease. The organisms *Chlamydia*, *Rickettsia*, *Ehrlichia*, and *Coxiella* are obligate intracellular pathogens, requiring host protection for survival.
- The mycobacteria, causative agents of tuberculosis and leprosy, also represent organisms that successfully survive inside host cell within phagocytic compartments. *Mycobacterium* spp. have a unique cell wall containing high quantities of long-chain mycolic acids, important in their pathogenesis. Other species containing mycolic acids that render them acid-fast are *Nocardia*, *Cryptosporidium*, and *Isospora* spp.

Self-assessment questions can be accessed at www.StudentConsult.com.

Basic Virology

13

CONTENTS
- VIRAL CLASSIFICATION AND STRUCTURE
- VIRAL GENETIC MATERIAL: RNA OR DNA
- STRATEGIES FOR INFECTIVITY AND REPLICATION
- VIRAL DISEASE PATTERNS AND PATHOGENESIS
- DIAGNOSTIC VIROLOGY
- THERAPY AND PROPHYLAXIS FOR VIRAL INFECTIONS

Viruses are small entities (20 to 300 nm) whose genomes replicate inside cells using host cellular machinery to create progeny virions (virus particles). On its own, a virus may be considered an inert biochemical complex of macromolecules because it cannot replicate outside a living cell. However, viruses are known to infect all living organisms, and a broad variety of viruses contribute to human disease.

KEY POINTS ABOUT VIRUSES
- Viruses are entities whose genomes replicate inside cells using host cellular machinery to create progeny virions (virus particles) that can transfer their genome to other cells.
- A broad variety of viruses are of high medical significance and contribute to manifestations of human disease.

VIRAL CLASSIFICATION AND STRUCTURE

Historically, viruses were named according to common pathogenic properties, organ tropism, and modes of transmission. Viral groups are classified based on viral host range, particle morphology, and genome type. Viral hosts represent species from all classes of cellular organisms; prokaryotes (including the Archaea and Bacteria), eukaryotes (including algae, plants, protozoa, and fungi), and complex invertebrates and vertebrates. Indeed, viruses can cross phyla; for example, different members of Poxviridae can infect vertebrates and insects.

Virion structure varies among different viral groups, yet all virus particles are enclosed by a capsid structure that surrounds the viral genome. Icosahedral symmetry is the preferred capsid morphology (Fig. 13-1). The capsid is a protein shell composed of repeating subunits, or protomers (also referred to as *capsomeres*). The capsid together with the enclosed nucleic acid is called the *nucleocapsid*. The term **virion** denotes the complete infective virus particle. Many viruses demonstrate the classic capsid polyhedron structure of 20 equilateral triangular faces and 12 vertices, which define axes of fivefold rotational symmetry. However, alternative virion morphologies exist. Some viruses have a helical nucleocapsid, consisting of a helical array of capsid proteins composed of identical protomers wrapped around a filament of nucleic acid. Thus, for these viruses, such as myxoviruses (e.g., tobacco mosaic virus), the length of the helical nucleocapsid is determined by the length of the nucleic acid.

Other virus families have an outer envelope consisting of a lipid bilayer surrounding the viral capsid. Such viral envelopes are derived in part from modified host cell membranes during particle formation and release (budding) from the infected cell. The exterior of the bilayer is studded with transmembrane proteins, revealed as glycoprotein spikes or knobs. Both the outer capsid and envelope proteins of viruses are glycosylated and are important in determining the host range and antigenic composition of the virion.

KEY POINTS ABOUT VIRAL CLASSIFICATION AND STRUCTURE
- Icosahedral symmetry is the preferred status for organization of virus structure subunits (protomers) making up the capsid protein shell.
- Other structural forms exist, yet they all share commonality in packaging of genetic material (DNA or RNA) for delivery to host cells upon infection.

VIRAL GENETIC MATERIAL: RNA OR DNA

Each virus carries within the protective capsid a nucleic acid–based blueprint for replication of infectious virus particles (virions). Once a virus has invaded a cell, it is able to direct the host cell machinery to synthesize new progeny. The viral genome may be composed of RNA or DNA, single or double stranded. Encoded proteins may be nonstructural, such as nucleic acid polymerases required for replication of genetic material, or structural (those proteins necessary for assembly of new infectious virions). However, all viruses lack the genetic information encoding proteins necessary to generate

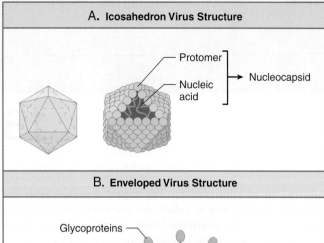

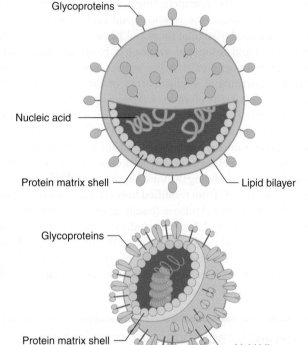

Figure 13-1. Basic virus structure. The basic virus structure is an icosahedron having 20 equilateral triangular faces and 12 vertices. Lines through opposite vertices define axes of fivefold rotational symmetry with structural features repeating five times within each complete rotation about any axis (**A**). A nucleocapsid contains DNA or RNA encapsulated within a capsid composed of protomer subunits. Enveloped virions (**B**) support an outer lipid bilayer studded with transmembrane glycoprotein spikes. Other forms of viral structure exist, such as those with helical nucleocapsid symmetry in which size of the virus is dictated by the length of nucleic acid core (not shown).

metabolic energy or protein synthesis. The viral genome (DNA or RNA) rarely codes for more than the few proteins necessary for replication or physical structure.

Viruses as a group are the only class of organisms with subspecies that keep RNA as their sole genetic material. Likewise, they are the only group of self-replicating organisms with subspecies that use single-stranded DNA as genomic content. Multiple forms of virus genomes are found in virions infecting human cells (Fig. 13-2).

STRATEGIES FOR INFECTIVITY AND REPLICATION

The first stage of viral infection and subsequent replication involves entry into the host cell (Fig. 13-3). The host cell phenotype has a great deal of influence on the strategy the virus uses to gain access; in turn, specific virus types may use different strategies to gain access to the same cell type. In general, the steps involve attachment and penetration, uncoating of the virus genome, and synthesis of early proteins (enzymes involved in viral replication), followed by synthesis of late proteins or structural components required for assembly and release of the infectious virion.

Viral entry into the cell is usually a passive reaction that does not require energy on the part of the virus. Naked viral particles may enter by membrane translocation, in which the entire virus crosses the cell border intact (pinocytosis). Alternatively, the naked particle binds to cell surface receptors and subsequent invagination occurs by either clathrin-mediated endocytosis (e.g., Adenoviridae) or other endocytic mechanisms such as interaction with caveolae or lipid rafts. A naked viral particle may also bind to the cell surface and inject genomic material into the host cell without complete cellular penetration of the invading virion. Enveloped viruses must enter host cells by using mechanisms of membrane fusion, either by receptor-mediated endocytosis or through fusion of viral and host membranes followed by injection of genomic material into the host cell cytoplasm (e.g., HIV-1 and HIV-2). In all cases, the critical component is the release of the viral genome from its protective capsid so that it can be transcribed to form new progeny virions.

In many cases, viral genetic material undergoes translation using host cellular machinery before viral genome replication (the exception being retroviruses and negative-sense RNA viruses). The first proteins generated are usually nonstructural DNA or RNA polymerases. Nucleic acid replication produces new viral genomes for incorporation into progeny virions. In general, DNA viruses replicate mainly in the nucleus and RNA viruses mainly in the cytoplasm, but there are exceptions (e.g., poxviruses contain DNA but replicate in the cytoplasm). Retroviruses are a special category of RNA viruses that require reverse transcription of their single-stranded RNA genome to a double-stranded DNA intermediate, which is then integrated into the host cell genome before viral replication can take place. Retroviruses not only encode a reverse transcriptase enzyme as part of the virion but also package it into newly formed virions.

The next set of proteins to be transcribed are structural in nature, including capsid protomers and scaffolding proteins that are required for assembly of the virion together with the newly replicated viral nucleic acid. Assembly of viral nucleocapsids can take place in either the nucleus (herpesvirus, adenovirus) or cytoplasm (poliovirus) or on the cell surface (influenza). Diagnostic inclusions, sometimes visible by light microscopy, are the result of virions accumulating at the sites of assembly. The final stage of replication results in the release of newly formed virions from the host cell. This may occur by budding from the cell surface (enveloped viruses) or via host cellular secretory pathways in which Golgi-derived vesicles

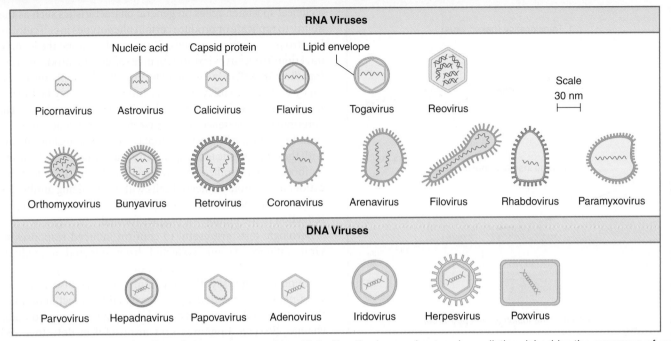

Figure 13-2. At least 21 families of viruses are capable of infecting the human host and are distinguished by the presence of an envelope or characteristic capsid and by internal nucleic acid genomic content.

are transported to the cell surface. In nonenveloped viruses, host cells are destroyed with consequent release of infectious virions upon cell lysis.

The Baltimore classification of viruses establishes seven groupings according to genome types and replication strategies (Table 13-1).

BIOCHEMISTRY

Clathrin-Mediated Endocytosis

Receptor-mediated endocytosis is a complex phenomenon in which binding of large extracellular molecules, such as viruses, to receptors on the cell membrane triggers assembly of clathrin triskelions. Specific clathrin adapter complexes are involved in transport across the membrane, resulting in endosomal compartment formation.

KEY POINTS ABOUT STRATEGIES FOR INFECTIVITY AND REPLICATION

- Initiation of viral replication begins with attachment and entry of viral particles into host cells, followed by replication of genetic material and production of assisting proteins (polymerases and structural proteins) required for assembly of virions with mature nucleocapsids.
- Newly formed virions are released from the host cells by assembly at the cell surface, via budding, or by lysis of the host cell.
- Multiple and diverse replication strategies are used depending on the type and class of genetic material contained within the virus entity.

VIRAL DISEASE PATTERNS AND PATHOGENESIS

Virus-induced pathology is the result of direct viral action leading to host cell death and tissue damage with subsequent sequelae. Almost all naked (nonenveloped) viruses produce acute infections in this manner as a result of cell lysis during replication and spread of infection to surrounding host cells. However, pathologic damage often is a result of an active immune response to viral antigens and epitopes presented on the surface of infected cells (Fig. 13-4). This becomes especially apparent in chronic infections when persistent virus production allows vigorous development of cellular (T helper cell [T_H] and cytotoxic T lymphocyte generation) and humoral (B cells with specific antiviral antibodies) responses. In the case of latent infections, short periods of active viral replication are kept in check by active immune responses, only to reemerge when immune system surveillance wanes.

KEY POINTS ABOUT VIRAL DISEASE PATTERNS AND PATHOGENESIS

- Pathology associated with viral infections is directly linked to viral cell tropism and mechanism of associated replication.
- Damage to surrounding tissue may be a direct result of the immune response and attempts to limit viral replication and spread.

DIAGNOSTIC VIROLOGY

The large number of possible viral agents that produce a given disease or pathology precludes the use of one simple test as diagnostic for a specific viral infection. Rather, laboratory

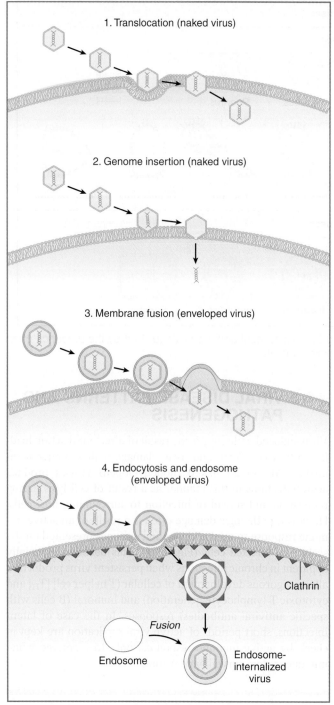

Figure 13-3. Viruses enter host cells following attachment by multiple means including (1) translocation, in which a virus crosses membranes intact; (2) genome insertion, in which attached viruses inject genetic material directly into cytoplasm; (3) membrane fusion, in which genomic contents of a virus are dumped into the host cell cytoplasm; and (4) endocytosis dictated by surface receptor binding and clathrin-mediated transport, sometimes leading to fusion into intracellular endosomes.

diagnosis is usually performed under the assumption of clinical disease spectrum, relying on symptoms and epidemiologic data (Fig. 13-5). Clinical observations alone are at times sufficient for diagnosis, allowing for clear therapeutic intervention prior to verification of virus identity. The identification of a virus from a clinical specimen relies on general characteristics such as the ability to replicate or produce certain phenotypes in cell culture. In vitro propagation in tissue culture can determine the level of infection. Typically, viral growth in tissue culture produces a cytopathogenic effect, which may be visualized as a plaque or vacancy within a monolayer of cells. Infected cells may be further fixed and stained; immunohistochemical methods can be used to determine the presence of characteristic and diagnostic inclusions (or inclusion bodies) in cytoplasm or in the nucleus. Along with these diagnostic tools, specificity may be obtained by using antibodies to neutralize the viral agent and confirm its identification. Incubation of infected cells with neutralizing antibodies will result in reduction of growth of the virus and subsequent reduction in the number of plaques formed.

Serologic tests are useful to confirm induced responses to viral antigens. Serum collected from infected individuals may be assessed for antibody response by the enzyme-linked immunosorbent assay. Enzyme immunoassays, also known as *solid-phase immunoassays*, are designed to detect antibodies through secondary production of an enzyme-triggered color change. Enzyme immunoassays are useful in detecting and quantifying the presence of viral agents in clinical specimens (blood, vaginal swabs, and feces). Use of specific antibodies immobilized to a solid phase also allows capture of pathogens for diagnostic quantitation to known standards.

Identification of viral agents in clinical specimens also may be performed by using highly sensitive genetic tools. RNA or DNA can be extracted and sequences probed by hybridization techniques. If the agent is already identified, analysis can also be accomplished by use of polymerase chain reaction or reverse-transcriptase polymerase chain reaction to specifically amplify sequences unique to the viral agent under consideration. However, in some instances, this task is difficult or impossible because of atypical clinical presentation or histopathologic features. Molecular diagnostics using DNA microarrays, or gene chip arrays, offer the promise of precise, objective, and systematic virus classification from clinically obtained specimens. In addition, diagnostic gene chip arrays that carry sequences of major clinically relevant viral pathogens allow extremely rapid screening of small diagnostic samples.

●●● THERAPY AND PROPHYLAXIS FOR VIRAL INFECTIONS

Therapy against viral infection makes use of chemotherapeutic agents that are effective to control infection and disease. Virucides, such as detergents, chloroform, and ultraviolet light, use general mechanisms to limit environmental spread of viruses. This is especially effective at reducing levels of enveloped organisms that are sensitive because of their bilipid envelope. Antiviral agents are more specific and include molecules that target receptors for cell attachment or that inhibit viral penetration, uncoating of the viral genome, or viral replication. Inhibition of replication may occur at multiple stages; agents may be directed against macromolecular synthesis and inhibit transcription, translation, or posttranslational modification (e.g., protease inhibitors). The majority of clinically

Therapy and prophylaxis for viral infections

TABLE 13-1. Different Classes of Viruses Grouped According to Replication Strategy (Baltimore Classification)

CLASS	GENOME TYPE	REPLICATION STRATEGY	EXAMPLES OF GENUS INFECTING HUMAN CELLS
I	dsDNA	Semiconservative method using bidirectional replication forks from a single origin	Poxvirus Herpesvirus Adenovirus Papovavirus
II	ssDNA	Formation of a replicative from double-stranded DNA intermediate	Parvovirus
III	dsRNA replicating via (+) RNA	Conservative mechanism in which input RNA is transcribed to mRNA	Reovirus
IV	ssRNA (+) sense genomes	Synthesis of (−) sense RNA on a (+) sense template	Coronavirus Flavivirus Astrovirus Picornavirus
V	ssRNA (−) sense genomes	Begins with transcription by virion-associated RNA-dependent RNA polymerase	Arenavirus Orthomyxovirus Paramyxovirus Rhabdovirus
VI	Diploid ssRNA	Uses a dsDNA, longer than genome length intermediate (provirus), which is integrated covalently into host cell chromosomal DNA	Retrovirus
VII	dsDNA	Uses gapped or nicked circular dsDNA genomes to replicate via longer than genome length messenger-sense ssRNA intermediates	Hepadnavirus

(+), Sense strand; (−), antisense strand.

available antiviral agents target nucleic acid synthesis and limit viral replication rather than completely eliminating organisms (virustatic vs. virucidal). Problems associated with host toxicity limit the use of virucidal agents in many instances. An additional concern is the selective pressure associated with prolonged use of antiviral agents, which allows mutants to arise that are no longer susceptible to drug action.

Another class of antiviral agents includes those that function as immunomodulators to improve host response to combat infection. These antivirals do not directly attack the specific pathogen, but rather they globally stimulate host immune responses. Interferons (IFNs) were first defined as glycoproteins that interfere with viral replication through degradation of viral mRNA. Most nucleated cells make IFN-α and IFN-β, secreted molecules that bind to specific receptors on adjacent cells to protect them against subsequent infection by progeny viruses. There are at least 17 different subtypes of IFN-α but only one subtype of IFN-β. In addition to direct antiviral effects, IFN-α and -β enhance expression of class I and class II major histocompatability complex molecules on infected cells, in effect increasing viral antigen presentation to specific T_H and cytotoxic T cells. A functionally related molecule, IFN-γ, produced by T_H1 cells, cytotoxic lymphocytes, and natural killer cells, is a potent activator of macrophages and a powerful antiviral immunomodulating agent.

A more specific approach is to synthesize antibodies that bind to viral pathogens to mark them for attack and clearance by other elements of the immune system. Vaccination (immunoprophylaxis) induces a primed state so that secondary exposure to a pathogen generates a rapid immune response, leading to accelerated elimination of the organism and protection against onset of clinical disease. Success depends on the generation of memory T and B cells and the presence in the serum of antivirus-specific neutralizing antibody. Neutralizing antibodies work to inhibit viral attachment, penetration, uncoating, and even viral replication. Alternatively, nonneutralizing antibodies can be quite therapeutically functional, assisting in viral clearance by marking the virus for phagocytosis by monocytes. Antiviral vaccines are effective when presented to the host in a manner that is similar to natural exposure to viral antigens, thus generating protective immunity. This may be accomplished through active immunization with live modified viral strains, with inactivated virus particles, and with subunit vaccines (Table 13-2). In general, active immunization leads to long-lasting immunity. Alternatively, passive antiviral treatment is possible by administration of high-titer, specific antivirus antibodies (hyperimmune globulin) that confer host

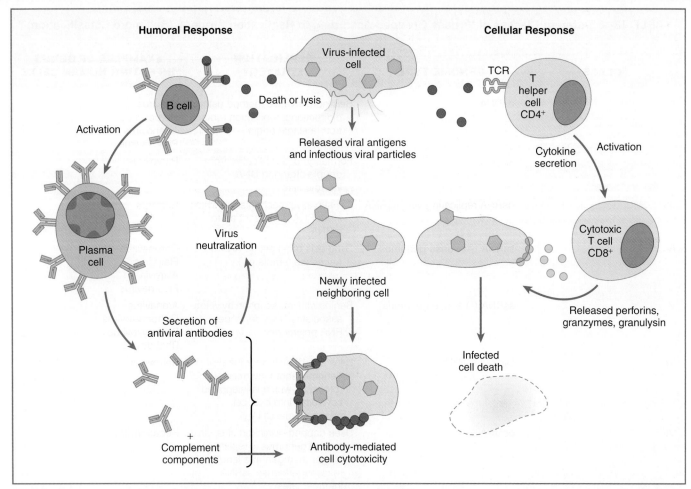

Figure 13-4. Damage to tissue may be initiated by response to released viral antigens and to viral antigens presented by major histocompatability complex molecules on the surface of infected host cells. Released antigens allow development of antibody responses (*left side*), leading to deposition on the surface of infected targets and cellular destruction by complement mediation. T helper cells release cytokines that assist cytotoxic lymphocytes to induce killing of target cells (*right side*). During both processes, bystander killing of surrounding tissue may occur, leading to manifestation of pathology. *TCR*, T-cell receptor.

resistance. Passive administration leads to fast-acting but temporary immunity to imminent or ongoing exposure.

PHARMACOLOGY

Antiviral Agents: Highly Active Antiretroviral Therapy

Highly active antiretroviral therapy holds great promise to limit HIV infection by means of agents that are terminal nucleoside analogs, non-nucleoside reverse transcriptase inhibitors, and viral protease inhibitors. Zidovudine was the first Food and Drug Administration approved antiretroviral and is an analog of thymidine that works as a nucleoside analog reverse transcriptase inhibitor.

KEY POINTS ABOUT THERAPY AND PROPHYLAXIS FOR VIRAL INFECTIONS

- INFs are a natural part of the immediate protective host response, which is elicited upon invasion by viruses.
- Among the few successful chemotherapeutic agents that inhibit viral replication are nucleoside analogs that compete with normal nucleotides for incorporation into viral DNA or RNA and protease inhibitors that interfere with virus assembly.
- Immunization with subunit vaccines or live attenuated viruses increases both antibody production and long-term protective cell-mediated responses directed at viruses upon subsequent infection.

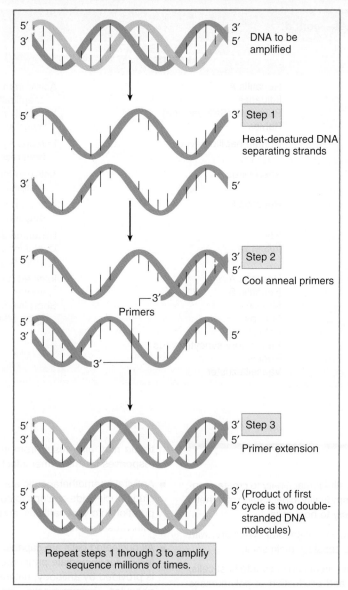

Figure 13-5. Viral identification can be accomplished by multiple means, including cell culture, microscopic identification, serologic methods to detect antiviral antibodies, direct detection of viral antigens, or detection of nucleic acids by polymerase chain reaction. The method of polymerase chain reaction amplification of viral nucleic acids is depicted here.

TABLE 13-2. Virus Vaccines

VACCINE CLASS	VIRUS	COMMENTS
Live virus vaccines	Adenovirus Chickenpox Measles Mumps Poliovirus (Sabin vaccine) Rotavirus Rubella Smallpox (variola) Yellow fever	Active immunization using avirulent attenuated strains Effective at inducing antibodies and cytotoxic lymphocyte responses

Continued

TABLE 13-2. Virus Vaccines—cont'd

VACCINE CLASS	VIRUS	COMMENTS
Killed virus vaccines	Hepatitis A Influenza Poliovirus (Salk vaccine) Rabies	Active immunization using heat or chemically inactive virus particles Vaccination may be combined with other viruses (polyvalent)
Virus-like particles	Human papillomavirus	Immunization using particles assembled from recombinant coat proteins
Subunit vaccines	Adenovirus	Active immunization using purified proteins
Polypeptide vaccines	Hepatitis B	Active immunization using synthesized polypeptide protein sequences
DNA vaccines (evaluation only)	HIV Influenza	Experimental Useful for induction of cytotoxic T-lymphocyte response
Passive antibodies	Hepatitis A Hepatitis B Measles Mumps Rabies Respiratory syncytial virus Rubella Varicella-zoster	Injected purified antibodies obtained from another source Short-lived function with little value when given after disease onset

HIV, human immunodeficiency virus.

KEY POINTS

- Viruses replicate by using host cellular machinery to create progeny virions. All viruses share commonality in packaging of genetic material (DNA or RNA) for delivery to host cells.
- Viruses are organized according to icosahedral symmetry, with protomer subunits comprising a capsid protein shell.
- Viral replication begins with attachment and entry into host cells, replication of genetic material, and production of polymerases and structural proteins required for production and subsequent assembly of virions with mature nucleocapsids.
- Newly formed virions are released from the host cells by assembly at the cell surface, via budding, or by lysis of the host cell.
- Viral cell tropism and mechanism of replication play a major role in development of associated pathology. Subsequent damage to tissue may also result from immune recognition and targeted responses to limit further infection of neighboring cells.
- Antiviral chemotherapeutic agents that inhibit viral replication include nucleoside analogs to compete with nucleotides for incorporation into viral particles, and protease inhibitors that interfere with virus assembly.
- All nucleated cells are capable of producing a subclass of interferons for immediate protective host responses; additional help is provided by antibody, natural killer cells, and adaptive T lymphocytes. Preimmunization increases both antibody production and long-term protective cell-mediated responses, allowing quicker and more effective responsiveness upon infection.

Self-assessment questions can be accessed at www.StudentConsult.com.

Clinical Virology 14

CONTENTS
- **RESPIRATORY VIRUSES**
  - Adenovirus
  - Influenza Virus
  - Parainfluenza Virus
  - Rhinovirus
  - Coronavirus
  - Respiratory Syncytial Virus
- **MUMPS, MEASLES, AND OTHER CHILDHOOD EXANTHEMS**
  - Mumps
  - Rubella
  - Rubeola
  - Parvovirus
  - Varicella
- **POXVIRUSES**
- **ENTEROVIRUSES**
  - Poliovirus
  - Coxsackievirus
  - Echoviruses
- **HEPATITIS VIRUSES**
- **HERPESVIRUSES**
  - Human Herpes Simplex Virus Types 1 and 2
  - Cytomegalovirus
  - Epstein-Barr Virus
  - Human Herpesviruses 7 and 8
- **RHABDOVIRUSES**
  - Lyssavirus
- **PAPOVAVIRUSES**
  - Papillomaviruses
  - JC and BK Viruses
- **RETROVIRUSES**
  - Human Immunodeficiency Virus
  - Human T Lymphocyte Virus
- **ARBOVIRUSES**
  - Dengue Virus
  - Yellow Fever
  - Japanese Encephalitis Virus
  - West Nile Virus
  - Tick-Borne Encephalitis
  - Vesiculovirus
  - Other Arboviruses
- **PRIONS**
- **EMERGING VIRAL PATHOGENS**

Viruses are entities that infect and replicate within host cells and are able to direct cellular machinery to synthesize new infectious particles. The extent of infection and associated pathology depends on the number of virions infecting the host as well as the physical damage and trauma associated with the infective process. Many times, host recognition of viral antigens and immune response contribute greatly to disease and pathologic manifestations. Selected viruses of clinical significance are listed in Table 14-1.

●●● RESPIRATORY VIRUSES

The main agents for clinical respiratory infections include the adenovirus, parainfluenza virus, respiratory syncytial virus (RSV), and rhinovirus. These viruses cause disease in the upper respiratory tract that leads to symptoms of pharyngitis (sore throat) with accompanying coryza (nasal discharge), tonsillitis, inflammation of the sinuses and middle ear, fever, and myalgia (muscle pain). Infection of the lower respiratory tract may induce bronchitis with inflammation of the larynx and trachea, exhausting cough and wheezing, and bronchopneumonia.

Adenovirus

The adenovirus group represents approximately 50 identified species of nonenveloped, double-stranded DNA viruses. The adenoviruses are frequently associated with asymptomatic respiratory tract infection that produces cellular cytolysis in the pharynx and subsequent host inflammation and cytokine response from immune T cells. Of interest, certain species easily lend themselves to genetic engineering and have the potential for gene therapy in clinical settings. During lytic infection, the adenovirus enters human epithelial cells, replicates, and causes host cell death. Transition to latent infection involves lymphoid tissue, where the virus may remain dormant for longer periods. There are reports that adenovirus infection may be linked to oncogenesis and cancer, although this has not been firmly established. Clinical symptoms include acute respiratory disease and pharyngeal inflammation with related fever; immunocompromised hosts are susceptible, and special care should be taken to monitor postoperative, therapeutically induced immunosuppression in transplant patients.

TABLE 14-1. Selected Viruses of Clinical Significance

VIRUS CLASS	VIRAL AGENT	CLINICAL MANIFESTATIONS
Respiratory virus	Parainfluenza Rhinovirus SARS RSV	Pharyngitis and coryza Tonsillitis and sinus inflammation Fever and myalgia
Childhood exanthem	Mumps Rubella virus and rubeola Erythema infectiosum Chickenpox	Exanthems Rashlike macules
Poxvirus	Smallpox variola virus	Papulovesicular lesions
Enterovirus	Poliovirus Coxsackievirus	Meningitis Exanthems and myocarditis
Hepatitis virus	Hepatitis A through E	Liver dysfunction, hepatocellular carcinomas
Herpesvirus	HSV-1, HSV-2, CMV, EBV, VZV Mononucleosis Lymphomas	Cold sores and fever blisters
Rhabdovirus	Rabies virus *Lyssavirus*	Acute CNS infection
Papovavirus	HPV	Cutaneous or mucosal lesions
Retrovirus	HIV-1 and HIV-2 HTLV	AIDS, leukemia
Arbovirus	Dengue virus Yellow fever virus Japanese encephalitis virus West Nile virus	Viral hemorrhagic fever Encephalitis and meningitis
Prion	Prions	Spongiform encephalopathy
Emerging viral agent	Ebola virus *Hantavirus*	Hemorrhagic fever Pulmonary distress

SARS, severe acute respiratory syndrome; *RSV*, respiratory syncytial virus; *HSV*, herpes simplex virus; *CMV*, cytomegalovirus; *EBV*, Epstein-Barr virus; *VZV*, varicella-zoster virus; *HPV*, human papillomavirus; *HIV*, human immunodeficiency virus; *HTLV*, human T lymphocyte virus; *CNS*, central nervous system; *AIDS*, aquired immunodeficiency syndrome.

Influenza Virus

Human flu-causing viruses belong to one of three major influenza-causing orthomyxoviruses; influenza A virus, influenza B virus, or influenza C virus. The more recently characterized pandemic H1N1/09 virus has been determined to be a subtype of influenza A virus. Influenza viruses are RNA viruses that replicate in the cytoplasm and require virally encoded enzymes. They have two specific protuberances, hemagglutinin and neuraminidase. There are 15 basic shapes of the hemagglutinin and 9 of neuraminidase, with nomenclature named accordingly. The spikes are hemagglutinin, which binds avidly to sialic acid residues on cells that work like grappling hooks. These viruses have segmented genomes, allowing them to form hybrid strains upon coinfection with host cells containing different viral strains. The resultant mix leads to a term called *antigenic drift* among surface proteins, which is especially seen in the hemagglutinins H1 and H2 of influenza virus A. The clinical symptoms of infection include fever, cough, sore throat, muscle aches, and conjunctivitis. In severe cases, fatal pneumonia may follow infection. Vaccination against influenza allows production of antibodies that can neutralize neuraminidase as well as block entry of viral attachment to target host cells.

Parainfluenza Virus

Parainfluenza viruses are paramyxoviruses that are the causative agents of nearly 40% of acute respiratory infections in infants and children. Human parainfluenzas are serotyped as paramyxoviruses 1 through 4; serotype 4 consists of subtypes A and B. The virus is acquired through inhalation of infected respiratory droplets, with the nasopharynx as the primary site of infection. The virus attaches to cell membranes by way of a hemagglutinin trimeric protein that binds cell surface glycoproteins with neuraminic acid residues. Clinical symptoms range from the simple common cold to croup, bronchitis, and bronchopneumonia. Symptoms are usually accompanied by a hoarse or "barking" cough, sometimes with a swollen epiglottis.

Rhinovirus

Rhinoviruses (among other viruses that infect the upper respiratory tract) cause the common cold or coldlike symptoms (discharging or blocked nasal passages, sneezing and rhinorrhea, and sore throat). There are more than 120 serotypes, which are primarily spread by aerosolization. The most common route of infection is the nose, and infections do not spread to the lower respiratory or intestinal tracts. Infection spread is limited by simple washing of hands. Most rhinoviruses replicate in epithelial cells of the nasal mucosa, eventually leading to edema of connective tissue. Cellular destruction is commonly caused by immune response to viral infection and antigens and not by direct viral cytotoxic activity.

Coronavirus

Coronaviridae are best known as the second most frequent cause of the common cold. Although coronaviruses are known to cause disease in birds and pigs, recent studies indicate that the agent responsible for the severe acute respiratory syndrome (SARS) belongs to this genus. SARS is a serious, life-threatening viral infection that is thought to have mutated to a human transmissible form. The mortality rate of SARS is high compared with other common respiratory viral infections, with up to 10% of fatalities due to respiratory failure after symptoms of hypoxia, cough, and dyspnea (labored breathing).

Respiratory Syncytial Virus

RSV, a paramyxovirus, is the main cause of bronchiolitis (inflammation of the bronchioles) and pneumonia in children. As the name implies, infection in tissue culture causes cells to form syncytia, resulting in large, multinucleated cells.

●●● MUMPS, MEASLES, AND OTHER CHILDHOOD EXANTHEMS

Childhood viral-related exanthems (eruptive disease) include rubeola (measles) and rubella (German measles), chickenpox, roseola infantum, and erythema infectiosum (a parvovirus; also called *fifth disease*). All these infections affect the respiratory system and are highly contagious. Unless immunocompromised, children with these illnesses usually recover fully without treatment. As with all viral infections, a heavy load may result in more severe complications, including pneumonia, heart and kidney damage, and encephalitis. In the United States and most of Europe, successful vaccination programs have limited the incidence of disease, although these viruses remain a major cause of childhood deaths in other regions of the globe. The measles, mumps, and rubella vaccine protects by using live, attenuated viral antigens.

Mumps

Mumps is a paramyxovirus that causes inflammation of the salivary, parotid, and submaxillary glands. In adults, orchitis and oophoritis (inflammation of the testis and ovaries) may occur. A common clinical association is that of aseptic meningitis due to disseminated infection after viral crossing of the blood-brain barrier.

Rubella

Rubella virus, which causes German measles, is the single member of the *Rubivirus* genus of the togavirus family. Infection results in rashlike macules on the face, body, and limbs. There is associated lymphadenopathy at the back of the head and neck, which may proceed to arthralgia (joint pain) and symptoms of arthritis as infection disseminates. Special care should be taken regarding rubella infections of women during the first trimester of pregnancy because spread of infection to the fetus leads to congenital abnormalities (e.g., heart defects, deafness, mental retardation) in the majority of cases.

Rubeola

Rubeola (measles) is a common childhood illness caused by a morbillivirus, resulting in manifestation of a diagnostically distinct exanthema. The virus is spread through direct contact with nasal or throat discharge from an infected individual. The disease is highly contagious and typically persists for 7 to 10 days. The hallmark of infection is a rash that usually begins on the face and spreads to the rest of the host's body, accompanied by fever and cough. After symptoms disappear, the individual is left with a lifelong immunity against subsequent infection.

Parvovirus

Erythema infectiosum caused by parvovirus B19 is similar to measles. It is also called *fifth disease* or *slapped-cheek disease*, the latter because of its tropism for immature red cells (erythrotropism). This leads to the characteristic facial sequela of a lacy, pink, macular rash on the cheeks. Because of its predilection for immature red blood cells, transmission from mother to fetus can be life-threatening for the unborn child.

Varicella

The causative agent for chickenpox (varicella) is a member of the human herpesvirus family. Infection occurs by inhalation with replication of the virus in the mucosa of the upper respiratory tract, followed by dissemination throughout regional lymph nodes. In the case of chickenpox, the resulting pathology manifests as a macule that develops to a pustular lesion, which eventually crusts over. The virus has the ability to establish latent infection in neurons that can reactivate when cellular immunity is impaired or wanes from advanced age. Varicella-zoster virus (VZV) causes herpes zoster (shingles), a herpetic neuropathology in which reactivated virus travels through axons to infect the dermatome supplied by the sensory ganglion, thus introducing painful vesicular lesions on the skin. Another related herpesvirus, human herpesvirus 6 (HHV6), is responsible for roseola infantum, causing a febrile illness with a rash in newborns.

●●● POXVIRUSES

Smallpox (variola) is an orthopox virus that has been virtually eliminated in humans as a result of international efforts to control disease through intense vaccination with cross-reactive vaccinia. Related pathogenic poxviruses in humans include those caused by cowpox and molluscum contagiosum, all of which cause papulovesicular lesions on the hands, forearms, or face. Monkeypox is a generalized infection that may be more invasive and involve lung tissue.

●●● ENTEROVIRUSES

Enteroviruses are members of the picornavirus family. They are orally or fecally transmissible; ingested viruses infect cells of the oral cavity and pharyngeal mucosa as well as draining lymphoid tissue (tonsil, cervical, and eventually mesenteric). After replication, they are shed into the alimentary tract and intestinal tissue. The enteroviruses include poliovirus, coxsackieviruses A and B, echovirus, and hepatitis A. The overall clinical symptoms are typically mild but occasionally serious disease persists with continued shedding of virions for months to years.

Poliovirus

Before the introduction of vaccines, paralytic poliomyelitis was a major health threat. Introduction of the Salk vaccine, which uses inactivated virions, and the Sabin vaccine, which uses attenuated virions, quickly led to high levels of protection in the general population. There are three serotypes of poliovirus (1, 2, and 3), with serotype 1 being the most prevalent. Infection with poliovirus can be life-threatening; within 1 week after infection the virus has the potential to disseminate, entering the blood supply and spreading to the anterior horn of the spinal cord and then to the motor cortex of the brain. Localized paralysis ensues, depending on which neurons are infected and the amount of damage caused during infection.

Coxsackievirus

Coxsackievirus types A and B can cause pathogenic infection. Coxsackie type A is typically associated with exanthems, whereas both types A and B can lead to myocarditis. If untreated, infection may result in aseptic meningitis involving headache and fever, stiff neck, and malaise. Coxsackie type A can cause painful mouth ulcers of the tongue and hand-foot-mouth disease, exhibited by blisters on the hands and feet. It may also cause Bornholm disease (sometimes called *devil's grip*), manifested as sharp chest and side pains (epidemic pleurodynia).

Echoviruses

Enteric cytopathic human orphan viruses (Echoviruses) cause symptoms similar to rhinovirus infection. They also may cause exanthems and myocarditis, similarly to the coxsackie viruses described. Rashes may be maculopapular, similar to rubella (German measles), or of a lighter, subcutaneous form (roseola-like, vesicular, or petechial).

●●● HEPATITIS VIRUSES

Five recognized hepatitis viruses are known to cause liver-related pathology. They are hepatitis A, B, C, D, and E. Hepatitis A and E are enterically transmitted, whereas the others are parenterally spread (Fig. 14-1). Hepatitis A (called *infectious hepatitis*) is caused by the picornavirus enterovirus 72; hepatitis E is a calicivirus. In both cases, the virus is spread by fecal-oral route, replicates in the intestinal tract before infecting hepatocytes, and eventually is spread through the stool.

Human hepatitis B (HBV) is a hepadnavirus, similar to strains found in the woodchuck and duck. The virus is spread by needles, through sexual encounters, and from mother to fetus during birth. HBV replicates in the liver, with a large load of virus particles shed during infection that are available for immune recognition. Antibodies to the surface antigen (HBsAg) are a useful diagnostic tool for infection detection. Persistent infection leads to chronic liver damage with accompanying cirrhosis and potential for development of hepatocellular carcinoma. A relatively new vaccine directed against the hepatitis B surface antigen is available and seems to be effective in the prevention of liver tumorgenic development. Hepatitis C is a flavivirus that also causes hepatocellular carcinomas. Hepatitis D, or "delta agent," is a defective virus that requires HBV to replicate, and it seems to exacerbate symptoms related to HBV. Finally, a related hepatitis G flavivirus has been identified but does not seem to be a major cause of liver disease.

●●● HERPESVIRUSES

The herpesviruses include eight members that cause pathology, including the VZV and HHV6 described above. In all cases, latent infection is established after a primary persistent and lytic indication, sometimes lying dormant throughout the lifetime of an individual. Table 14-2 lists properties of the herpesviruses.

Human Herpes Simplex Virus Types 1 and 2

Human herpes simplex virus types 1 and 2 (HHSV1 and HHSV2) are characterized by infectious virions that begin with cytopathic infections that progress to latent infection of neurons and ganglia (Fig. 14-2). In many cases, reactivation occurs under conditions of stress and immunosuppression, or even unrelated immune hypersensitivity reactions. Pathology of vesicular lesions occurs via both direct cytolytic activity and immune response attacking infected tissue. Common manifestations include cold sores, or fever blisters, which may often be seen around the mouth and lips. Of interest, almost all adults have been exposed to HHSV1 at some time,

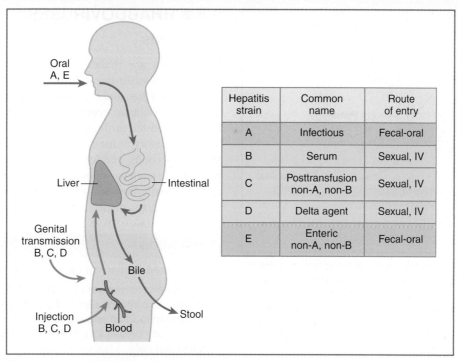

Figure 14-1. Routes of transmission for the hepatitis viruses.

TABLE 14-2. Properties of Human Herpesviruses

VIRUS STRAIN	COMMON NAMES	CLINICAL MANIFESTATION	FEATURES
Human herpes simplex virus 1	HHSV1, HSV-1	Cold sores or fever blisters (oral, keratitis)	Cytopathic progression to latent infection of neurons and ganglia
Human herpes simplex virus 2	HHSV2, HSV-2	Painful genital blisters	Similar to HHSV1
Human cytomegalovirus	HCMV	Fever and lymphocytosis	Mononucleosis-like illness
Epstein-Barr virus	EBV	Lymphadenopathy, sore throat, hepatosplenomegaly	Persistent in B lymphocytes (Burkitt lymphoma)
Varicella-zoster virus	VZV	Chickenpox and shingles	Latent infection in neurons
Human herpesvirus 6	HHV6	Roseola infantum	T-cell lymphotropic virus
Human herpesvirus 7	HHSV7, HHV7	No known diseases	"Orphan" virus
Human herpesvirus 8	HHSV8, HHV8	Primary effusion lymphoma	Considered agent responsible for Kaposi sarcoma and Castleman disease

with positive antibody titers present in serum. HHSV2 is the causative agent for genital herpes.

> **NEUROSCIENCE**
>
> **Efferent Motoneurons**
>
> Somatic efferent motoneurons originate in the spinal cord and synapse with muscle fibers to direct contraction of skeletal muscle. These motoneurons are asymmetric units composed of a dendrite, a soma, and axons, which may be several feet in length. These neurons may be exploited by viruses (such as herpes or varicella) where retrograde transmission allows axonal transport of an infectious agent to other regions of the body.

Cytomegalovirus

Human cytomegalovirus (HCMV) infects most individuals within the first few years of childhood, with approximately 70% of adults having immunoglobulin (Ig) G antibodies specific for cytomegalovirus antigens. The virus is harbored in CD4+ and CD8+ cell phenotypes. The virus is secreted in saliva; infection may result in a mononucleosis-like illness associated with hepatitis, fever, and lymphocytosis (increase in blood lymphocytes). Infection of healthy individuals does not lead to symptoms; however, patients who have undergone organ transplantation and immunosuppressed individuals are at much greater risk for development of infection-related pathologies.

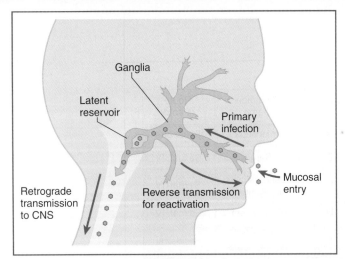

Figure 14-2. Herpes simplex virus establishes a primary infection of the skin or oral mucosa and spreads to nearby ganglia, where it may remain latent. Upon reactivation, reverse travel of virus in a ganglion allows spread to skin or mucosa, causing occurrence of surface lesions. Retrograde transmission may occur as well, leading to meningitis upon infection of the central nervous system (CNS).

> **PATHOLOGY**
> **Burkitt Lymphoma**
>
> Burkitt lymphoma is a high-grade B-cell lymphoma associated with EBV. The basis of the pathology is thought to occur through infection and transformation of B cells, leading to malignancy through chromosomal translocation events. The most common translocation moves the protooncogene *Myc*-containing segment of chromosome 8 to chromosome 14q band 32, placing it close to the immunoglobulin heavy chain gene which results in Myc overexpression.

Epstein-Barr Virus

Epstein-Barr virus (EBV) is similar to HCMV in that nearly all adults have antibodies specific for EBV antigens. The virus is persistent in B lymphocytes and is the agent responsible for infectious mononucleosis and X-linked lymphoproliferative syndrome. Infectious mononucleosis ("kissing disease") is characterized by lymphadenopathy, sore throat, and hepatosplenomegaly; because of the predilection of B cells for infection, atypical lymphocytes are present in peripheral blood. When reactivated, EBV can cause Burkitt lymphoma and other carcinomas.

Human Herpesviruses 7 and 8

Human herpesvirus 7 (HHSV7) is an "orphan" virus; it has been identified and isolated; no known diseases have been attributed to it, but recent evidence suggests it may be associated with exanthematic disease (roseola), or linked to central nervous system seizures. Human herpesvirus 8 (HHSV8) is thought to be the agent responsible for the Kaposi sarcoma that is characteristic of later stage human immunodeficiency virus (HIV) infection.

RHABDOVIRUSES

Lyssavirus

Lyssaviruses are rhabdoviruses. The genus *Lyssavirus* includes the neurotropic rabies virus, the Mokola virus, the Duvenhage virus, and the European and Australian bat lyssaviruses. The classic madness associated with rabies virus is well documented through history and popular literature. Dogs serve as a reservoir for disease, as do coyotes, wolves, jackals, skunks, raccoons, and foxes. Shortly after infection, the virus migrates through peripheral nerves and enters the brain, causing acute pathology within the central nervous system, leading to neurologic disorders and sometimes death.

> **HISTOLOGY**
> **Keratinized Epidermal Layers of the Skin**
>
> An outer protective layer of keratinized stratified squamous epithelium covers the dermis. Keratinocytes, primarily within the stratum lucidum and stratum corneum, produce a fibrous scleroprotein that is the chief structural component of hair and nails.

PAPOVAVIRUSES

Papillomaviruses

Human papillomaviruses (HPVs) are papovaviruses that are clinically defined according to their ability to induce cutaneous or mucosal lesions (Table 14-3). Mechanistically, these viruses infect basal epidermal keratinocytes, where they lay dormant until epidermal cells migrate to the skin surface.

TABLE 14-3. Papillomavirus Infections

CUTANEOUS PAPILLOMA	ANATOMIC SITE
Common wart (verruca vulgaris)	Hands
Plantar wart	Feet
Butcher's wart	Hand
Flat wart	Arms, face, knees
Extragenital Bowen disease	Extremities, head
Epidermodysplasia verruciformis	Exposed skin
Mucosal papilloma	Mucosa
Laryngeal	Larynx
Conjunctival	Arms, face, knees
Oral	Mouth
Anogenital papilloma	Genital, anal areas
Condyloma acuminatum	Genital, anal areas
Cervical intraepithelial neoplasia	Neck
Bowen disease	Groin
Buschke-Löwenstein tumor	Glans penis, perianal area
Vulvar intraepithelial neoplasia	Genital area
Penile/anal intraepithelial neoplasia	Genital area

This induces the virus to replicate, altering the appearance of host tissue and resulting in cutaneous or mucosal localized verrucae (warts), which appear as hyperkeratotic nodules. Papillomaviruses have anatomic preferences, with some causing lesions on the skin surface and others causing genital or anal lesions. In addition, there is strong evidence that certain papillomaviruses are linked to cervical and genital cancers. HPV vaccines are now available that function to limit development of genital warts and cervical cancers.

JC and BK Viruses

Two additional members of the papovavirus family are known to infect humans. The JC virus replicates in neural tissue and is associated with progressive, multifocal leukocyte encephalopathy. The BK virus has been isolated from the urinary tract, indicative of the kidney being the site of infection. BK virus may cause interstitial nephritis and cystitis. Neither causes major pathologic damage unless the patient is immunocompromised or immunosuppressed.

RETROVIRUSES

Retroviruses require reverse transcription of their single-stranded RNA genome to a double-stranded DNA intermediate for integration into the host cell genome. The retroviruses represent infectious agents that may be oncogenic or cytolytic. This group includes the slow-acting lentiviruses of which acquired immunodeficiency syndrome (AIDS)-associated HIV-1 and HIV-2 are members, having high clinical importance and enormous social and economic impact throughout the world.

Human Immunodeficiency Virus

HIV-1 is the retrovirus most commonly associated with HIV infection in the United States and Europe. HIV-2, which has high nucleotide sequence homology to HIV-1, is prevalent in regions of Africa. Both are lentiviruses that infect CD4+ cells, including T helper lymphocytes, macrophages, and other cell types. The gp120 virus surface antigen binds with high affinity to the cell surface CD4, promoting fusion between virus and cell membranes. Coreceptor chemokine receptors CXCR4 and CCR5 may also play a role in internalization as well as immunoglobulins that assist in antibody-dependent uptake (Fig. 14-3). The natural history of HIV depicts the outcome of HIV infection, which leads to destruction of lymphocytes and development of AIDS (Fig. 14-4).

Human T Lymphocyte Virus

The human T lymphocyte virus (HTLV) is an oncogenic retrovirus transmitted by human contact and associated with adult T-cell leukemia. The different subtypes of HTLV are classified according to their ability to infect T lymphocytes. Presentation of infection may be asymptomatic, may give rise to neurologic disorders, or can result in lymphoproliferative events. HTLV-l infection is the main type associated with acute adult T-cell leukemia.

ARBOVIRUSES

Arboviruses are primarily transmitted by arthropod vectors, including those viruses spread by mosquitoes and ticks. The virus has a reservoir in another species (zoonotic), with humans being a secondary host for the infective agent.

Dengue Virus

Dengue virus is a flavivirus with four related subtypes. The virus is spread by a bite of the mosquito *Aedes aegypti* or *Aedes albopictus*. The major pathology during infection is due to increased capillary permeability, which leads to subsequent pleural effusion, ascites, and thrombocytopenia. Petechial skin hemorrhages may occur as well as gastrointestinal bleeding.

Yellow Fever

Yellow fever is another flavivirus that causes viral hemorrhagic fever. It is commonly found in Africa and South America and is spread by multiple mosquito species. The disease may lead to liver failure, renal insufficiency, and, in some cases, proteinuria (excess protein in urine). The jaundice associated with the disease led to its common name.

Japanese Encephalitis Virus

Japanese encephalitis virus is a mosquito-borne (*Culex tritaeniorhynchus*) flavivirus found in East and Southeast Asia. Domestic pigs and wild birds serve as reservoirs. The disease manifests as acute encephalitis with rigor, including neck stiffness, muscular weakness, and partial paralysis, and cachexia (wasting).

West Nile Virus

West Nile virus causes a viral-induced encephalitis and meningitis. The virus is endemic in the Near East, Africa, and Southwest Asia; it has recently been found in North America. Wild birds transmit the virus by way of multiple mosquito species (*Culex, Aedes,* and *Anopheles*). The virus crosses the blood-brain barrier and infects brain parenchyma and the leptomeninges.

Tick-Borne Encephalitis

Tick-borne encephalitis is yet another flavivirus that causes disease involving the central nervous system. The disease manifests as meningitis, encephalitis, or meningoencephalitis. The reservoir is the *Ixodes* tick, and the disease is found in Russia and East Asia.

Vesiculovirus

The vesicular stomatitis virus is also a rhabdovirus. It is thought that infection may be transmitted by sandflies and mosquitoes, with humans being infected directly by the transcutaneous route.

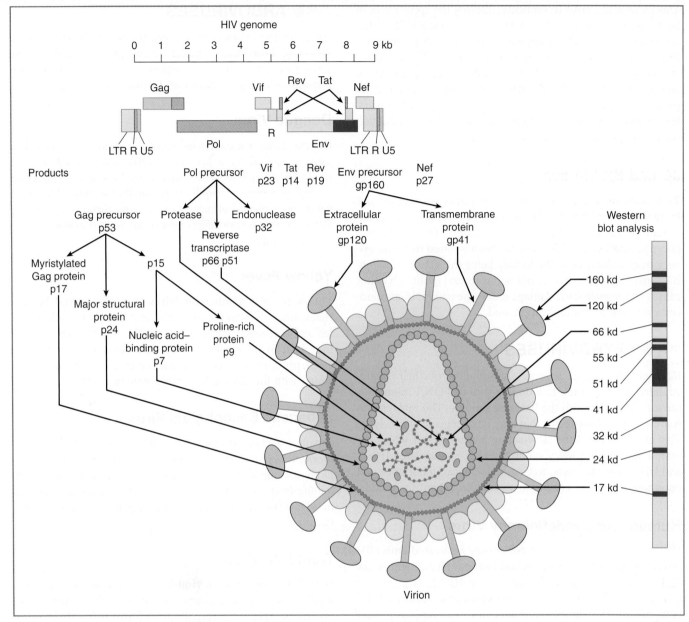

Figure 14-3. The human immunodeficiency virus (HIV) retrovirus is composed of an outer envelope glycoprotein coat surrounding a lipid bilayer. Internal group antigens provide core protein and scaffold structure. Internal reverse transcriptase and regulatory proteins interact with viral RNA, allowing virus gene expression and protein assembly for replicating virions.

Other Arboviruses

Ross River virus is an alphavirus in the togavirus family. It is transmitted via mosquitoes and causes polyarthritis and acute fever, with the majority of cases reported in Australia and the South Pacific islands. The Syndbis virus is another alphavirus member of the Togaviridae, spread by mosquitos and causing polyarthritis as well as arthralgias. Rift valley fever is also spread by mosquitoes and is caused by *Phlebovirus* of the Bunyaviridae family.

●●● PRIONS

Prion disease is a form of spongiform encephalopathy once believed to be of viral origin. The etiology is related to a prion protein that has the same primary amino acid sequence as a host protein found at low levels in healthy brain tissue. The prion protein has an abnormal conformation and triggers conversion of the normal host cellular isoform to an infectious form. Accumulations of the abnormal form are deposited as long filamentous aggregates of amyloid within the brain that gradually damage neuronal tissue. The diseases associated with this infection are Creutzfeldt-Jakob disease, Gerstmann-Straussler disease, and kuru.

●●● EMERGING VIRAL PATHOGENS

Emerging viral pathogens are viruses that suddenly appear in the environment with severe economic and social impact. These agents are usually extremely virulent and often are originally zoonotic. The species-jumping infectious agent finds a

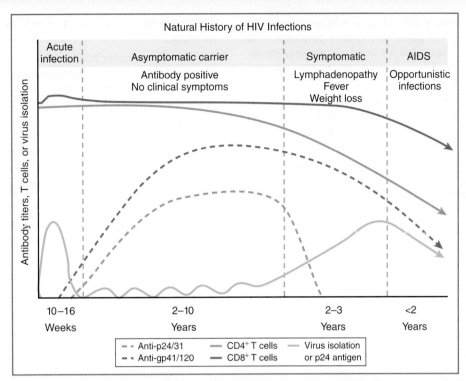

Figure 14-4. Infection of CD4+ lymphocytes (and other cell types) leads to virus production and cytolysis or long-term latent infection that progresses from primary infection through late symptomatic infection (*AIDS*). Accompanying this process are profound defects in T helper and cytotoxic cell activity, with concomitant development of opportunistic infections. *HIV*, human immunodeficiency virus.

TABLE 14-4. Some Emerging Viral Pathogens

VIRAL AGENT	FAMILY	INFECTION	ZOONOTIC RESERVOIR
Ebola virus	Filoviridae	Hemorrhagic fever	Possibly primates
Marburg virus	Filoviridae	Hemorrhagic fever	Possibly African green monkey
Lassa virus	Arenaviridae	Hemorrhagic fever	Rodents
African swine fever virus	Iridoviridae (formerly)		Warthog + tick vector
Morbillivirus, equine (Nipah virus)	Paramyxoviridae	Encephalitis, respiratory disease	Horse
Hantavirus	Bunyaviridae	Pulmonary distress	Rodents
Lyssavirus, bat	Rhabdoviridae	Rabies-like illness	Bat
Severe acute respiratory syndrome	Coronaviridae	Pulmonary distress	Birds

suitable host in humans, either through mutation or adaptation. A list of emerging viral pathogens is given in Table 14-4.

KEY POINTS

- The main agents for clinical respiratory infections include the adenovirus, parainfluenza virus, influenza virus, RSV, and rhinovirus. Infection leads to clinical symptoms of pharyngitis, coryza, inflammation of the sinuses and middle ear, fever, and myalgia.
- The childhood viral-related exanthems are highly contagious and produce "eruptive disease." They include rubeola (measles) and rubella (German measles), chickenpox (VZV), roseola infantum, and erythema infectiosum.
- Enteroviruses are orally or fecally transmissible members of the picornavirus family, infecting the oral cavity and pharyngeal mucosa. The enteroviruses include poliovirus, coxsackieviruses A and B, echovirus, and hepatitis A.
- Hepatitis viruses are spread by oral or genital transmission. Hepatitis A and E are spread by fecal-oral routes, whereas hepatitis B, C, and D are sexually or intravenously spread. All can cause chronic hepatic infection with subsequent liver damage.
- The herpesviruses include eight members that cause pathology. HHSV1 and HHSV2 establish primary infection in the skin or oral cavity and progress to latent infection of neurons and ganglia. The infection begins as cytopathic; retrograde transmission may

Continued

lead to meningitis. Nearly 70% of all individuals have been exposed to HCMV. The Epstein-Barr virus is persistent in B lymphocytes and is responsible for infectious mononucleosis. HHSV8 is thought to be responsible for Kaposi sarcoma development.

- HPVs are within the papovavirus designation and are clinically defined according to their ability to induce cutaneous or mucosal lesions. New studies link infection with development of cervical cancers.

- Retroviruses require reverse transcription of their single-stranded RNA genome to a double-stranded DNA intermediate prior to host cell genome integration. These include HTLV as well as the slow-acting lentiviruses such as HIV-1, which is most commonly associated with HIV infection. The HIV-1 virus is linked to destruction of CD4+ lymphocytes and development of AIDS.

- Arboviruses are transmitted by arthropod vectors. The virus has a zoonotic reservoir with humans being a secondary host for the infective agent. Examples of these include the dengue viruses, the Japanese encephalitis virus, and the West Nile virus.

- Emerging viral pathogens are usually extremely virulent and often are originally zoonotic. The species-jumping infectious agent finds a suitable host in humans, either through mutation or adaptation.

Self-assessment questions can be accessed at www.StudentConsult.com.

Mycology   15

CONTENTS
- **BASIC MYCOLOGY**
  - Fungal Structure and Classification
  - Fungal Pathogenesis
  - Antifungal Agents
  - Laboratory Diagnosis of Fungal Infections
  - Immune Response to Fungal Agents
- **CLINICAL MYCOLOGY**
  - Superficial and Cutaneous Mycoses
  - Subcutaneous Mycoses
  - Systemic Mycoses

●●● BASIC MYCOLOGY

Fungi are filamentous, spore-producing, eukaryotic organisms that do not contain chlorophyll. More than 100,000 species grow as saprophytes (requiring organic material for energy), some with beneficial use in production of foods such as cheese, wine, and beer. However, nearly 300 species are pathogenic for humans. Mycotic species are known to cause allergic reactions (to molds and spores), elicit toxic effects (mycotoxicosis) upon ingestion of fungal products, and lead to infection whose severity is determined by the immune status of the host.

Fungal Structure and Classification

The mycotic species include molds, yeasts, and higher fungi. The phyla of medically important fungi are listed in Table 15-1. Fungi have a complex and rigid structural cell wall made primarily of the polysaccharides chitin, glucans, and mannans. The membrane is bilayered with inclusion of sterol components (ergosterol and zymosterol), in contrast to the cholesterol typically found in higher eukaryotic membranes. Fungi are usually aerobic and exhibit both sexual and asexual forms of reproduction.

Almost all fungi form filaments called *hyphae*, which are the filamentous subunits of molds and mushrooms. They may exist as septate forms (with cross-walls) or as aseptate coenocytic forms (without cross-walls). A colony or mass of multibranched intertwined hyphae is referred to as a *mycelium;* a fungal body may also be called a *thallus*. Yeasts are single-celled forms of fungi that generally reproduce by budding. Under differing environmental conditions, dimorphic fungi can change morphology between yeast and hyphal structures. Some fungi produce spores, formed asexually or sexually, which may contribute to allergic responses in certain individuals. Others reproduce through fragmentation. The different forms are depicted in Figure 15-1.

Fungal Pathogenesis

Fungi do not produce endotoxins or exotoxins. Rather, infection in many cases is due to colonization of the organism at local or systemic tissue sites. Low pH, fatty acids, and rapid turnover of skin surface epithelial cell layers limit infections. Antibodies (primarily immunoglobulin [Ig] A with some assistance by IgG) at mucosal surfaces may also contribute by limiting growth and spread of fungal agents.

Agents may be classified according to tissue colonized. Superficial mycoses are limited to the skin and hair. For example, tinea nigra and tinea versicolor are superficial skin infections giving rise to macular lesions. Black piedra and white piedra colonize hair shafts, resulting in brittle hair. Cutaneous mycoses include infections of keratinized tissue, usually hair and nails. Dermatophytes, such as those causing ringworm of the scalp (tinea capitis), folliculitis (tinea barbae), and athlete's foot (tinea pedis), are included in this category. Subcutaneous mycoses, usually introduced by trauma or breach of dermal tissue, involve infections of subcutaneous fascia, muscle, and deeper epidermal layers. Chromomycosis caused by soil fungi results in abscesses, with lesion development along lymphatics as infection spreads. Mycetomas and sporotrichosis infection may also cause cutaneous lesions represented as pustular nodules as organisms are cleared from subcutaneous sites. All three (superficial, subcutaneous, and cutaneous mycoses) are typically noninvasive infections. Systemic mycoses, on the other hand, can be life threatening, especially when the host is immunocompromised. Although normal immune responses leave the host asymptomatic, opportunistic disease may ensue if an individual is immunosuppressed. Systemic infections result from inhalation of pathogenic fungal spores, followed by differentiation and dissemination to other tissue sites. Histoplasmosis, coccidiomycosis, and blastomycosis are three dimorphic fungi present as mold forms in soil and yeast in tissue. Although usually self-limiting, dissemination to other tissue sites may occur, with ensuing destructive lesion development.

Mycotoxicoses are caused by ingestion of fungal toxins in contaminated foods. Ergotism, caused by ergot alkaloids from *Claviceps purpurea* released into grains, leads to pronounced vascular and neurologic disorders due to inhibition of epinephrine. These alkaloids include lysergic acid

TABLE 15-1. Phyla of Medically Important Fungi

PHYLA	SPORE TYPE	CHARACTERISTICS	REPRODUCTION
Ascomycota	Ascospores	Penicillium mold, yeast, *Aspergillus* aflatoxin	Sexual reproduction in a sac called an *ascus*
Deuteromycota (mitosporic fungi)	Imperfect fungus	Ringworm, athlete's foot	No formation of sexual spores
Oomycota	Zoospores	Aquatic "water molds"	Asexual spore reproduction
Zygomycota	Zygospores, sporangiospores	Bread molds, rice fermenters	Sexual reproduction by gametes as well as asexual reproduction
Basidiomycota	Basidiospores	Causative agents of plant diseases	Sexual reproduction in a sac called a *basidium*

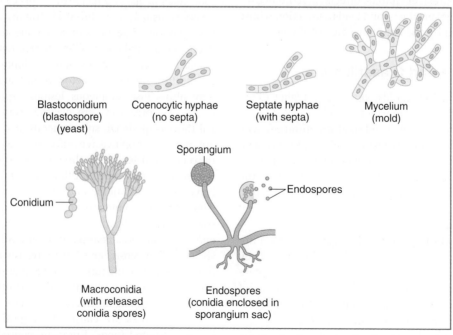

Figure 15-1. The growth of fungi generally involves two phases: vegetative and reproductive. In the vegetative phase, the cells are haploid and divide mitotically. Most fungi exist as molds with hyphae, but some fungi exist as unicellular yeast cells. Some fungi can change their morphology and are termed *dimorphic*. In the reproductive phase, fungi undergo either asexual or sexual reproduction. Asexual reproduction involves the generation of spores.

diethylamide (LSD), a psychotropic agent. Additional psychotropic toxins include those released from *Psilocybe*. Ingestion of *Amanita* mushrooms and associated amanitine and phalloidin toxins can lead to liver damage. *Aspergillus flavus* aflatoxins, found in contaminated grains and peanuts, are bisfuranocoumarin metabolites (coumarin derivatives) that cause hepatic hemorrhage and sever liver damage.

Allergic reactions to fungal spores are dependent upon the host's susceptibility, combining deleterious hypersensitive and antibody IgE-mediated responses. In many cases, immediate asthmatic reactions occur, with associated bronchoconstriction due to antibody-mediated release of histamines from mast cells and eosinophils.

membrane sterols are particularly effective (nystatin, amphotericin B), as are azole derivatives that inhibit ergosterol biosynthesis (miconazole, fluconazole, ketoconazole). Grisans (griseofulvin) may be given orally to treat more invasive dermatophytes, and work to inhibit microtubular function (Fig. 15-2).

PHARMACOLOGY

Lysergic Acid Diethylamide: Psychotropic Agent

LSD is a psychotropic agent synthesized from an alkaloid found in the fungus ergot (*Claviceps purpurea*). LSD is related structurally to several other drugs (e.g., bufotenine, psilocybin, harmine, ibogaine) and intensifies sense perceptions. The mechanisms of action include blocking of serotonin function (the indole amine transmitter of nerve impulses) in brain tissue.

PHYSIOLOGY

Pathophysiology of Allergen-Induced Bronchoconstriction

Fungal allergens can induce transient IgE-mediated syndrome with bronchoconstriction due to airway inflammation and edema and mucus hypersecretion, with subsequent changes in the distribution of ventilation. Allergens recognized by mast cells and basophils via high-affinity IgE receptors (FcεRIs) trigger immediate release of preformed mediators such as histamine, enzymes, hydrolases, and proteoglycans. Lipid mediators, such as leukotrienes and prostaglandins, are released in a secondary wave minutes later to exacerbate bronchoconstriction of smooth muscle and initiate mucus secretion.

Antifungal Agents

Because the cell wall consists primarily of chitin, fungal agents are not susceptible to common antibiotics that inhibit peptidoglycan synthesis. Effective therapy uses a combination of proper delivery to the site of infection (superficial vs. systemic) and directed activity against fungus-specific targets. For example, superficial infection of the skin or hair may be treated by site-specific cleansing in combination with topical applications. Subcutaneous or systemic infections require more aggressive and prolonged treatment. Targets against

Laboratory Diagnosis of Fungal Infections

Superficial and cutaneous mycoses may be identified by microscopic analysis, either by analysis of wet skin scrapings or by visualization after staining. For visualization, treatment of skin with alkaline solution dissolves tissue, leaving resistant material available for further staining and identification by morphologic analysis. Identification by culture is possible, but because of the ubiquitous nature of fungal entities, the laboratory procedures must take into account the possibility of airborne contamination. For culture procedures,

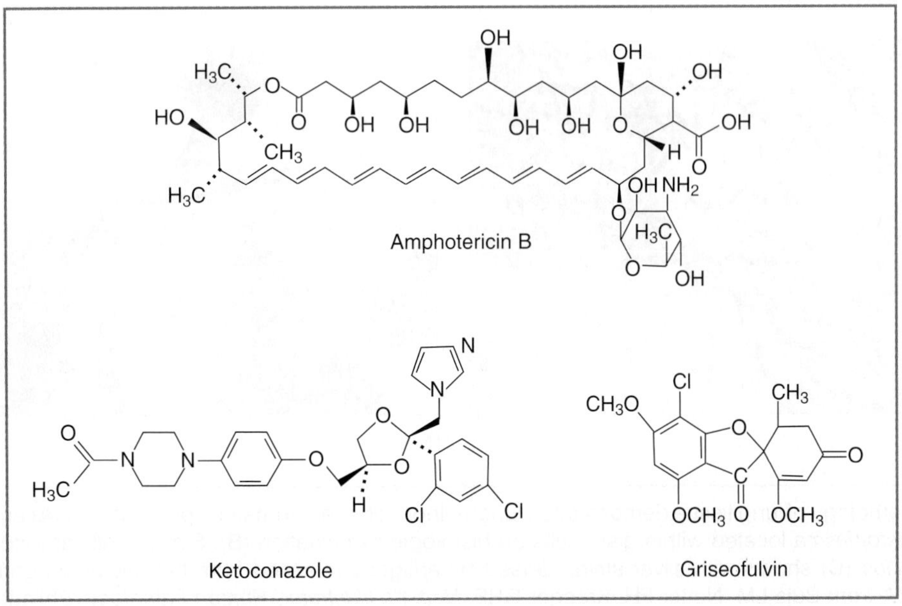

Figure 15-2. Molecular structure of selected antifungal agents.

slow-growing organisms are cultured on Sabouraud agar, which typically contains high levels of antibiotics to inhibit bacterial growth. More recent methods of identification include polymerase chain reaction analysis for specific DNA sequences and enzyme-lined immunosorbent assay analysis to confirm fungal antigens. Diagnosis of dermatophytes requires recovery of organisms from lesions. Systemic mycoses are usually identified serologically and immunologically, with confirmation by culture and polymerase chain reaction performed on material isolated from infected tissue.

Immune Response to Fungal Agents

The major immune responses to fungal agents are delayed-type hypersensitivity and granulomatous responses (Fig. 15-3). However, not much is known about immune response to fungal agents; rather, disease and pathology more often than not ensue when there is a failure in one arm of immunity. Cellular immunity appears to be the most important immunologic factor in resistance to fungal infections, although humoral responses (antibodies) may play a role. The importance of cellular reactions is indicated by the intense mononuclear infiltrate and granulomatous reactions that occur in tissues infected with fungi and by the fact that fungal infections are frequently associated with depressed immune reactivity of the delayed type (opportunistic infections). Chronic mucocutaneous candidiasis refers to persistent or recurrent infection by *Candida albicans* of mucous membranes, nails, and skin. Patients with this disease generally have a form of immune deviation (i.e., a depression of cellular immune reactions), with high levels of humoral antibody. Fungi appear to be resistant to the effects of antibody, so that cell-mediated immunity is needed for effective resistance.

When systemic mycoses give rise to clinical manifestations, the initial reaction manifests as an accumulation of neutrophils around the postcapillary venules. This reaction subsides and is replaced by mononuclear cells, consisting of T lymphocytes and macrophages, which locate around venules in a perivascular cuff. The endothelial cells that line venules become porous to blood proteins, leading to fibrinogen leakage into surrounding tissue. Fibrinogen is subsequently converted to fibrin. The accumulation of mononuclear cells plus fibrin causes induration at the site of exposure, the hallmark of delayed-type hypersensitivity.

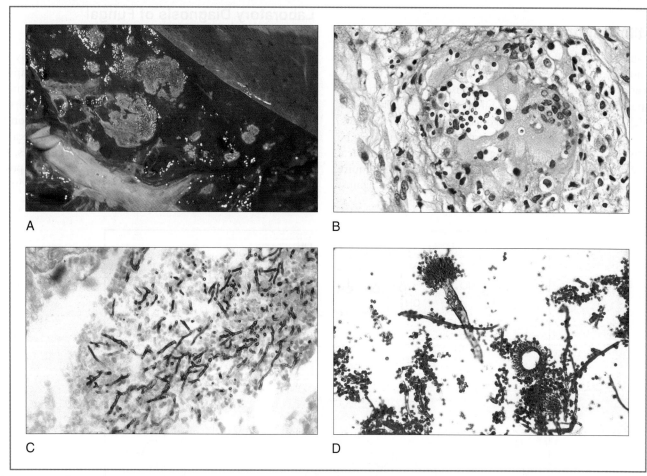

Figure 15-3. Gross pathology of lung tissue demonstrates fungus-induced granulomatous pneumonitis (**A**) and associated periodic acid-Schiff–stained *Histoplasma* located within giant cells on histologic examination (**B**). Similar cellular response is evident upon infection with *Aspergillus* (**C**) shown with silver stain, caused by antigens released from the etiologic agent isolated from sinus maxillary tissue (**D**). (**A** from Buja LM, Netter FH, Krueger GRF: *Netter's Illustrated Human Pathology*. Philadelphia: WB Saunders, 2004.)

CLINICAL MYCOLOGY

Superficial and Cutaneous Mycoses

The superficial mycoses include those that superficially infect skin or hair (Fig. 15-4). *Malassezia furfur* (tinea versicolor, pityriasis versicolor) infection of the skin causes hypopigmented or hyperpigmented macular lesions along the trunk of the body. *Exophiala werneckii* (tinea nigra) infection is usually asymptomatic and superficial, with hypopigmented areas appearing on the palms or on tanned skin. Associated lesions are usually flat and dark as a result of infection of stratum corneum layers of the epidermis. Piedra is a general term for fungal infection of hair shafts. *Piedraia hortae* (black piedra) infection results in hard black deposits along the hair shafts, while *Trichosporon beigelii* (white piedra) forms white deposits similar to a collar-like sleeve around hair follicles.

Cutaneous mycoses of the skin, hair, or nails are caused by dermatophytes belonging to three genera: *Epidermophyton*, *Microsporum*, and *Trichophyton*. Some instances may also be defined by *Candida* (yeast) infection. *Trichophyton rubrum*, *Trichophyton mentagrophytes*, and *Epidermophyton floccosum* are causative agents of athlete's foot, resulting in chronic scaly tinea pedis (scales on soles and heels). These same agents, along with *Candida*, may also cause acute infection of the groin (tinea cruris, jock itch), often accompanied by athlete's foot. Although less common, these same agents, along with *Microsporum canis*, may infect skinfolds of the anus (tinea corporis), leading to annular lesions that are pustular. *Trichophyton verrucosum* is the agent implicated in folliculitis of the facial hair (tinea barbae).

Ringworm of the scalp skin and hair (tinea capitis) is also considered a cutaneous mycosis, characterized as anthropophilic (epidemic, *Microsporum audouinii*) or zoophilic (nonepidemic, *Microsporum canis* or *Trichophyton mentagrophytes*). Anthropophilic tinea capitis is characterized by noninflammatory, grayish patches of hair and may be spread by contact with infected headgear. Zoophilic tinea capitis is typically of greater inflammatory nature, with resulting baldness (alopecia). *Trichophyton tonsurans* tinea capitis is very common in children and causes infections within the hair (endothrix). *Trichophyton schoenleinii* (tinea favus) is a more aggressive disease, with scalp scarring and hair loss occurring if it is not treated in a timely manner.

Candida spp. (discussed below) are normal fungal flora inhabitants of the intestines. However, it should be noted that some species, such as *Candida albicans*, may cause dermatomycoses and ensuing skin lesions (e.g., vulvovaginitis), otomycosis (infection of the auditory canal), and onychomycosis (infection of the fingernails). Thrush, infection of the tongue and inner mouth, is prevalent in immunocompromised individuals, especially those with late-stage HIV infection.

Three genera of actinomycetes were thought to be fungi but have more recently been shown to be bacteria that are related to mycobacteria. Two of these are discussed in Chapter 12: *Actinomyces* and acid-fast *Nocardia*. The third group, *Streptomyces*, causes a subcutaneous disease known as *fungus*

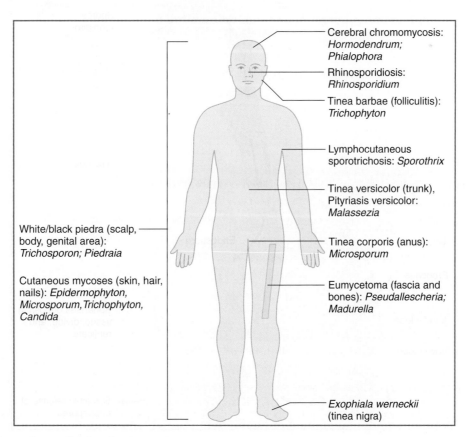

Figure 15-4. Locations of superficial and cutaneous mycoses.

tumor (mycetoma) that, if left untreated, can penetrate deep into tissue and infect bone material. The streptomycetes are aerobic like *Nocardia*.

Subcutaneous Mycoses

Three common subcutaneous mycoses are those causing sporotrichosis, chromomycosis, and mycetoma. All represent disease caused by saprophytic (soil-growing) fungi that enter tissue, usually through trauma. *Sporothrix schenckii* is the causative agent of lymphocutaneous sporotrichosis and is the most common form of subcutaneous nodular fungal disease. Clinical presentation demonstrates lesions following lymphatic vessels, with ulceration. Although lesions may be chronic, there is only minimal instance of systemic spread. Chromomycosis (or chromoblastomycosis) caused by dematiaceous (black-brown) fungi can appear as subcutaneous lesions that manifest along lymphatics. In many cases, such as verrucous dermatitis, granulomatous response ensues, leading to histologic presence of giant cells and lymphocytes in infected tissue. Cerebral chromomycosis manifests as focal brain lesions, with encapsulated masses present. Mycetoma manifests as deeper tissue lesions. Eumycotic mycetoma, caused by filamentous *Pseudallescheria boydii* and *Madurella grisea*, are organisms that typically drain into sinus tracts with associated swelling of dermal tissue. Other subcutaneous mycoses include rhinosporidiosis (*Rhinosporidium seeberi*), which is a chronic granulomatous disease of mucocutaneous tissue.

> **HISTOLOGY**
>
> **Lymphatics**
>
> Lymphatics are blind-ended, endothelium-lined tubes present in most tissues in similar numbers to capillaries. Lymphatics drain into collecting lymph nodes; in acute inflammation, the lymphatic channels become dilated and drain away fluid (inflammatory exudate), thereby limiting the extent of tissue edema.

Systemic Mycoses

Systemic mycoses are virulent diseases that are acquired most often through inhalation although they can be acquired by deep trauma. In many cases, they are referred to as *fungal pneumonias*. Fungi are usually dimorphic, with saprobic forms (deriving energy from organic material in soil) being filamentous and pathogenic forms more characteristic of small, unicellular yeasts. Inhaled spores induce primary responses in lung tissue; symptoms are typically of short duration (Fig. 15-5). Secondary infections are more invasive, with

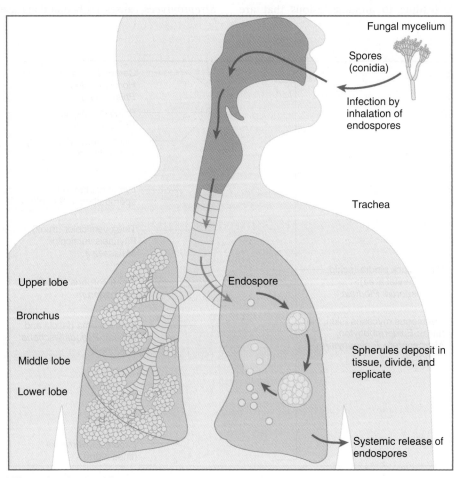

Figure 15-5. General life cycle of systemic mycoses.

potential spread to other organ systems. The main agents include the pathogens *Histoplasma*, *Coccidioides*, and *Blastomyces*; additional pathogenic organisms are briefly discussed.

Histoplasma

Inhalation of *Histoplasma capsulatum* hyphae fragments causes histoplasmosis. Saprobic hyphae contain microconidia, and tuberculate macroconidia are readily found in soils with high nitrogen content, such as when contaminated with bird droppings or bat guano. Upon inhalation, small budding yeast forms are phagocytosed, leading to infection of macrophages and histocytes. Clinically, patients may experience loss of weight and night sweats. A transient spread of disease by infected cells may occur, giving rise to flulike symptoms. Chronic infection and dissemination of organisms occurs within individuals who have immune defects, leading to prominent hepatosplenomegaly and lesion formation in sites distant from lung tissue.

> **PATHOLOGY**
>
> **Hepatosplenomegaly**
>
> Systemic infections with fungi such as *Histoplasma* in immunocompromised individuals can lead to development of hepatosplenomegaly. The systemic inflammation results in activation of the reticuloendothelial system with proliferation of tissue macrophages (histiocytes), both with and without intracellular organisms in the spleen and liver. This also leads to thrombocytopenia, pancytopenia, and coagulopathy.

Blastomyces

Blastomyces dermatitidis is the causative agent of blastomycosis. As a mold, it has characteristic hyphae and microconidia. Upon entry to host lung tissue, it appears as a yeast with a broad-based bud. Primary lung infection is limited in most individuals, controlled via macrophage phagocytosis and subsequent elimination. However, progressive pulmonary disease may occur, with further dissemination to skin or bone, characterized by granulomatous response.

Coccidioides

Coccidioidomycosis, caused by *Coccidioides immitis* and *Coccidioides posadasii*, usually presents an flulike illness with fever, cough, headaches, rash, and myalgias. Some patients do not recover and develop chronic pulmonary infection or widespread disseminated infection (affecting meninges, soft tissues, joints, and bone). Both *C. immitis* and related *C. posadasii* are thermally dimorphic fungi found in soil. Large, round, thick-walled spherules containing endospores are the typical structures formed in infected tissues.

Paracoccidioides

Paracoccidioides brasiliensis is the causative agent of a true systemic (endemic) mycosis called *paracoccidioidomycosis*. The spectrum of the disease is wide, varying from an asymptomatic infection verified by the skin test to a subclinical, symptomatic, or chronic infection. It is a thermally dimorphic fungus, growing as a mold in soil and as yeast in the host. In addition to the primary pulmonary form of the disease, acute pulmonary, chronic pulmonary, and disseminated forms may also be observed. In cases of disseminated paracoccidioidomycosis, the reticuloendothelial system, skin, and mucous membranes are frequently involved.

Candida

Candida is a yeast and the most common cause of opportunistic mycoses worldwide. It is a frequent colonizer of human skin and mucous membranes and one of the normal flora of skin, mouth, vagina, and stool. Although *Candida albicans* is the most abundant and pathologically significant species among the more than 150 identified, *C. tropicalis*, *C. glabrata*, *C. parapsilosis*, *C. krusei*, and *C. lusitaniae* also cause *Candida* infections (candidiasis). Disseminated infections arise by hematogenous spread from a primary infected locus. *C. albicans* adheres to host tissues, producing secretory aspartyl proteases and phospholipase enzymes that contribute to its pathogenicity. Immune recognition of glycans present in the cell wall is critical to initiate monocytic responses that eventually drive development of adaptive immunity. Infection may be acquired from exogenous sources (such as catheters or prosthetic devices) or by direct transfer, as seen in oral candidiasis in neonates of mothers with vaginal candidiasis.

Cryptococcus

Cryptococcus is an encapsulated yeast, with more than 50 identified subspecies. *Cryptococcus neoformans* is the causative agent of cryptococcosis, the most common clinical form of which leads to meningoencephalitis. The course of the infection is usually subacute or chronic. Cryptococcosis may also involve the skin, lungs, prostate gland, urinary tract, eyes, myocardium, bones, and joints. Four serotypes have been identified, of which three have been renamed as *Cryptococcus gattii*. The polysaccharide capsule and phenol oxidase enzyme are considered major virulence factors. Infection begins with inhalation of the yeast form. It is thought that melanizing enzymes from the agent protect the fungal cell from oxidative stress after being engulfed by phagocytic cells. Although natural killer cells participate in the early killing of cryptococci as do antibody-dependent cell-mediated mechanisms, the progression in human immunodeficiency virus–positive individuals confirms the requirement for effective adaptive functions to control infections.

Aspergillus

Aspergillus fumigatus and other *Aspergillus* spp. cause aspergillosis. Aspergillosis produces a variety of symptoms; bronchopulmonary aspergillosis caused by *A. fumigatus* causes a severe allergic reaction. Tissue macrophages and neutrophils play a critical role in containment and release fungicidal mediators. If not controlled, an aspergillum (a mycelial ball in the lung cavity) may form, causing hemoptysis (coughing up blood). Invasive aspergillosis found in immunocompromised persons results in the fungus growing outward from the lungs, invading blood vessels, and spreading to other organs. Spread

Figure 15-6. The aflatoxin B_1 is produced by the fungus *Aspergillus flavus*, typically found growing on peanuts. Aflatoxin B_1 is an aromatic amine derivative that is an extremely potent dietary carcinogen having mutagenic effects on hepatocytes.

of organisms may also occur when immune responses are inappropriate, such as when T helper 2 lymphocyte responses occur. Allergic reactions to fungal antigens are also common. Aflatoxins, a group of related mycotoxins, such as those produced by *A. flavus* or *A. parasiticus* can lead to severe liver damage (Fig. 15-6).

Mucor

The *Mucor* fungi cause the group of infections referred to as zygomycosis (mucormycosis). *Mucor* is a filamentous fungus found in soil, plants, and decaying fruits. The genus has several species, the more common ones being *Mucor amphibiorum*, *M. circinelloides*, *M. hiemalis*, *M. indicus*, *M. racemosus*, and *M. ramosissimus*. Zygomycosis includes mucocutaneous and rhinocerebral infections as well as septic arthritis, dialysis-associated peritonitis, renal infections, gastritis, and pulmonary infections. Clinical complications arise as a result of vascular invasion that causes necrosis of the infected tissue and perineural invasion. Morphologically, the organisms are seen as hyphae, sporangiophores, sporangia, and spores.

Pneumocystis

Pneumocystis carinii pneumonia, is a condition that occurs frequently as an opportunistic infection in individuals infected with human immunodeficiency virus. However, it can often be seen in premature babies and elderly individuals whose immune systems are not fully competent. Symptomatically, infection of lung tissue leads to filling of alveoli with fluids, thus preventing oxygen/gas exchange. Recent sequence analysis of rRNA classifies *P. carinii* as a fungus, moving it out of its historical placement as a protozoan. The causative agent was renamed *Pneumocystis jiroveci*.

KEY POINTS

- The mycotic species are filamentous, spore-producing eukaryotes that include molds, yeasts, and higher fungi.
- The mycotic species may be pathogenic, causing allergic reactions or manifestations of diseases induced by mycotoxins.
- Agents are classified into four major groups according to the tissue colonized.
- Superficial mycoses are limited to the skin and hair.
- Cutaneous mycoses are infections of keratinized tissue and include the dermatophytes.
- Subcutaneous mycoses involve infections of subcutaneous fascia, muscle, and deeper epidermal layers.
- Systemic mycoses, usually resulting from inhalation of spores, may be extremely severe in immunocompromised hosts.

Self-assessment questions can be accessed at www.StudentConsult.com.

Parasitology   16

CONTENTS
PARASITE PATHOGENESIS
PROTOZOA
  Intestinal Protozoa
  Extraintestinal Protozoa
NEMATODES
  Pinworms and Whipworms
  Ascaris
  Hookworm
  Strongyloides
  Trichinella
  Filariae
CESTODES AND TREMATODES
  Taenia and *Echinococcus*
  Schistosomes
LICE AND MITES
LABORATORY DIAGNOSIS OF PARASITIC INFECTIONS
IMMUNE RESPONSE TO PARASITIC INFECTIONS

●●● PARASITE PATHOGENESIS

The saying "dead host, dead pathogen" is exemplified in parasitology. By definition, parasites are multicellular pathogenic organisms that shelter within (or on) other organisms and derive benefit from their host. Evolutionary pressures have contributed to establishment of highly complex interactions, allowing relatively long periods of host survival with limited pathology so that parasitic life cycles may continue. However, relationships may not always be commensal (mutual parasitism) and may cause diseases such as amebic dysentery, malaria, schistosomiasis, and elephantiasis. These parasites include the protozoa and the helminths. The protozoa exist as both intestinal and extraintestinal species. Helminths are multicellular invertebrate worms that infect humans with resultant mild to severe pathology. The helminths include nematodes, cestodes, and flukes (trematodes), which exhibit various pathologic etiologies.

Parasites interact with hosts in many ways. Some organisms are considered symbionts and spend all their life primarily associated with a single host (symbiosis). An obligate parasite is one that can live only in association with its host, whereas a facultative pathogen also can survive in free form. Endoparasites are capable of living inside the host's body; those that exist on the host body surface are termed ectoparasites. Mutualism is defined as a specialized interaction in which survival depends on physiologic reactions provided by both host and parasite. Hosts may also be classified according to the role they play in the parasite life cycle. A definitive host is one for which sexual maturity is reached during association, whereas intermediate hosts serve only as temporary reservoirs for physical change (metamorphosis) to another stage. Finally, many pathogens exist within arthropod (or other invertebrate) vectors, which serve as an essential component for completion of the parasite life cycle. Parasites of medical importance are listed in Box 16-1.

Almost all interaction with parasites causes histopathology, the form of which usually depends on the numbers of organisms present. In heavy infection, necrosis may occur with death of tissue from persistent immune reactivity due to large antigenic loads and assault on parasites. In lesser infections, cellular and tissue swelling may occur, with associated increase in cell size due to the presence of protein-filled granules. Fatty degeneration may be apparent, with subsequent deposition of large amounts of fat within cells and surrounding tissue. In most cases, infection leads to cellular change, be it hyperplasia or hypertrophy as a result of parasitic stimulation or metaplasia and neoplasia as the cells surrounding parasites are transformed into phenotypes not normally present at sites of infection.

KEY POINTS ABOUT PARASITES
- Parasites are organisms that shelter within (or on) other organisms, deriving benefit from the host.
- Parasitic organisms may be facultative or require obligate interactions with the host.
- A definitive host is one in which sexual maturity is reached during association.
- Intermediate hosts (or vectors) serve only as temporary reservoirs, allowing physical metamorphosis to reach another stage.

●●● PROTOZOA

The protozoa are eukaryotic organisms that are either single celled or colonial. All have mechanisms of locomotion, whether via flagella (Fig. 16-1), allowing transport in fluid (blood or lymph), or through pseudopodia, as seen in

Box 16-1. MEDICALLY IMPORTANT PARASITES

Intestinal Protozoa

Balantidium coli
Blastocystis hominis
Chilomastix mesnili
Cryptosporidium spp.
Cyclospora spp.
Dientamoeba fragilis
Eimeria tenella
Entamoeba coli
Entamoeba histolytica
Endolimax nana
Giardia lamblia
Iodamoeba bütschlii
Isospora belli
Microsporidium spp.
Trichomonas hominis

Nematodes

Ancylostoma duodenale
Ascaris lumbricoides
Enterobius vermicularis
Necator americanus
Strongyloides stercoralis
Toxocara canis
Trichinella spiralis
Trichuris trichiura

Extraintestinal Protozoa

Acanthamoeba spp.
Babesia microti
Entamoeba gingivalis
Eimeria stiedae
Giardia duodenalis

Histomonas meleagridis
Leishmania spp.
Naegleria fowleri
Plasmodium spp.
Toxoplasma gondii
Trichomonas vaginalis
Trypanosoma spp.

Cestodes and Trematodes

Clonorchis sinensis
Diphyllobothrium latum
Echinococcus granulosus
Echinostoma revolutum
Fasciola hepatica
Fasciolopsis buski
Hymenolepis spp.
Moniezia expansa
Paragonimus westermani
Schistosoma spp.
Taenia spp.

Filariae

Brugia malayi
Dirofilaria immitis
Loa loa
Onchocerca volvulus
Wuchereria bancrofti

Lice and Mites

Dermatophagoides spp.
Eutrombicula alfreddugesi
Pediculus humanus
Phthirus pubis
Sarcoptes scabiei

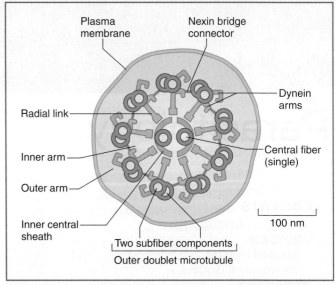

Figure 16-1. A flagellum (pl. *flagella*) is an extension of ectoplasm that moves in a whiplike manner to provide locomotion for unicellular protozoa. The basic structure is that of longitudinal cilium composed of microtubules arranged in nine doublets with two complete doublets in the center.

ANATOMY

Reproduction of Higher Eukaryotes

Reproduction of complex eukaryotes involves the union of two unlike sex cells (gametes), the sperm and the ovum, followed by the joining of their nuclei to form a zygote. This begins a rapid series of cell divisions (mitosis) in which segmentation occurs, giving rise to a hollow ball one cell thick, called a blastula.

amoebae. Pseudopodia (filopodia, axopodia, rhizopodia, and lobopodia) allow mobility along substrates but, unlike the flagella, do not assist in swimming within fluids.

The protozoa have nuclei as well as defined mitochondria and Golgi complex components. Indeed, certain protozoa possess acid hydrolase–filled lysosomes that assist in degradation of phagocytosed food particles. Common means of reproduction include asexual binary fission, as seen in amoebae, flagellates, and ciliates. Alternatively, schizogony (multiple fission) can occur, as seen with *Plasmodium* (malaria) in which multinucleated cells appear (schizonts) followed by breaking off of single-cell merozoites. In many cases (such as with merozoites) asexual replication may continue, or organisms may undergo further transformation to differentiated gametocytes for sexual reproduction. General clinical diagnostic tools for protozoan identification include the use of differential stains, such as iron hematoxylin or trichrome stains, or the Wright-Giemsa stain.

Intestinal Protozoa

The intestinal protozoa are aptly named because they cause the clinical complication of dysentery due to their residence in the lower gastrointestinal tract of the host. The common infective form is a cyst, usually transmitted by ingestion of contaminated food or water. In general, cysts modify their structure inside the host (excystation) to become trophozoites, which are the reproductive and feeding forms of the organism.

Amebae

Entamoeba histolytica is the major pathogenic intestinal protozoan and the cause of amebic dysentery in humans. It is a nonpathogenic commensal that is transmitted via a cyst stage through contact with infected feces or foods. Many other related amebae belonging to this group are nonpathogenic; for example, *Entamoeba coli* is a commensal that is commonly mistaken for *E. histolytica*. A general life cycle is depicted in Figure 16-2.

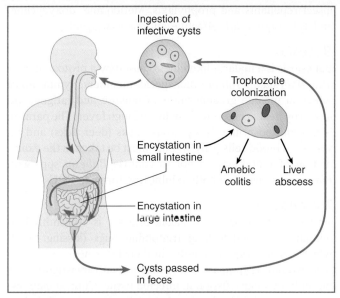

Figure 16-2. Life cycle of *Entamoeba histolytica*.

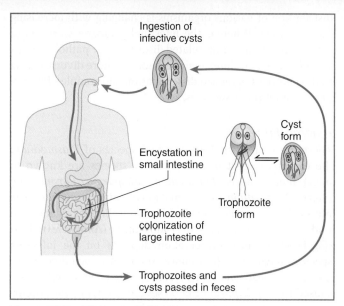

Figure 16-3. Life cycle of *Giardia*.

> **HISTOLOGY**
>
> **Histologic Organization of the Colon**
>
> The main function of the colon is fluid absorption and concentration of solid waste. The inner tunica mucosa contains specialized epithelial cells and glands (goblet cells) and a thin muscularis mucosae underlying the lamina propria. A submucosal layer with lymphatic nodules is surrounded by a muscularis externa, comprising outer longitudinal and inner circular muscle layers. The entire organ is contained by an overlying serosa.

Ciliates

Balantidium coli is a pathogenic ciliate that is the largest of the parasitic protozoans and is responsible for balantidial dysentery. Infection is acquired through ingestion of cysts from fecally contaminated water. *Balantidium* is partially invasive, causing intestinal lesions along submucosal tissue. Cyst forms excystate into trophozoites that multiply by binary fission.

Flagellates

Intestinal flagellates include pathogenic *Giardia lamblia* and *Dientamoeba fragilis*, as well as multiple nonpathogenic species. *Giardia*, one of the causative agents of traveler's diarrhea, and *Dientamoeba* are transmitted via infected food or placement of contaminated fingers in the oral cavity. The life cycle of *Giardia* is similar to that of amebae (described above), with human ingestion of infective multinucleated cysts derived from feces, and excystation to a trophozoite form (Fig. 16-3). *Dientamoeba* has a double-nucleated trophozoite stage but no known cyst form.

Coccidia

Cryptosporidium parvum, *Cyclospora* spp., and *Isospora* spp. are coccidian parasites that cause self-limiting diarrhea. The incidence of reported infections has increased in recent years, especially in immunocompromised individuals.

The general life cycle begins with the definitive host ingesting oocysts containing eight sporozoites. Once inside the intestine, developing merozoites within the schizont form are released to enter intestinal epithelium and begin the sexual cycle phase. Sexual union (gametogony) produces an oocyst, with oocytes being subsequently released to the environment. Because of intermittent findings in stool samples, positive diagnosis requires examination of three fecal specimens collected on alternate days. The oocysts of *Cyclospora* (8–10 μm) are roughly twice the size of *Cryptosporidium* oocytes (4 μm). In immunocompromised patients, *Cryptosporidium* may also inhabit respiratory and biliary tract tissue. *Isospora belli* is similar to *Cryptosporidium* in its life cycle, transmission, and symptoms. *Isospora* are usually diagnosed by the presence of cysts in feces using the same modified acid-fast stain as for *Cryptosporidium*.

The cat is thought to be the definitive host for *Toxoplasma gondii*, in which it resides as an intestinal parasite. In humans, *T. gondii* is an extraintestinal coccidian parasite that infects and undergoes schizogony in nucleated cells. *Toxoplasma* divides mitotically in humans as tachyzoites, giving rise to groups of organisms (pseudocysts or bradyzoites) that may reside in tissue sites (brain, heart, or skeletal muscles). Bradyzoite clusters may eventually become encysted, coinciding with developing immune responses. Toxoplasmosis is a common opportunistic infection associated with acquired immunodeficiency syndrome; weakened immune systems allow organisms to emerge from cysts to cause further tissue damage. Pregnant women who are carrying their first child are warned to limit exposure to cats to reduce the potential for infection and harm to the fetus.

Extraintestinal Protozoa

Extraintestinal protozoa include many organisms that are spread by vector agents. Many parasitic diseases are found throughout the tropical regions of the world; however, the distribution of various organisms is changing with increasing human travel. Whereas the intestinal protozoa were major contributors to intestine-related disorders, the pathologies associated with extraintestinal protozoa are more diverse owing to the nature of organism dissemination and complex interactions at multiple tissue sites.

Plasmodium

There are more than 150 identified species of *Plasmodium*, the causative agent of malaria. Four species are known to infect human hosts: *Plasmodium falciparum*, *Plasmodium vivax*, *Plasmodium ovale*, and *Plasmodium malariae*. The life cycle is complex and includes morphologic changes in both the human and the mosquito (*Anopheles*) host (Fig.16-4). Clinical presentation depends on the infecting species; however, all produce symptoms including fever and chills, headache and myalgias, and anemia. Infections caused by *P. falciparum* can be severe and fatal, with occurrence of central nervous system (CNS) inflammation (cerebral malaria) and accompanying acute renal failure. Travelers to endemic areas should limit exposure and use insect repellants and prophylaxis. Mefloquine, doxycycline, and chloroquine are effective antimalarial agents.

Babesia

Babesiosis is a malaria-like illness caused by a protozoan parasite *Babesia microti* or other species members. Symptoms of infection include gradual onset of malaise and fatigue, followed by headaches, muscle pain, and high fever. The parasite is transmitted primarily by *Ixodes* ticks (deer ticks) and invades red blood cells (RBCs). Care must be taken in the examination of blood smears not to confuse *Babesia* spp. with *Plasmodium* because both exhibit ring forms in RBCs.

Trypanosoma

Trypanosoma cruzi, or Chagas disease, is transmitted to humans by blood-sucking triatomine bugs (kissing bugs), which release trypomastigotes in their feces near the blood-meal wound. Trypomastigotes transform into amastigotes that multiply by binary fission in infected tissue. Clinical manifestations begin with a local lesion at the site of inoculation, with accompanying inflammation and edema. Self-resolving acute-stage symptoms include low fever, lymphadenopathy,

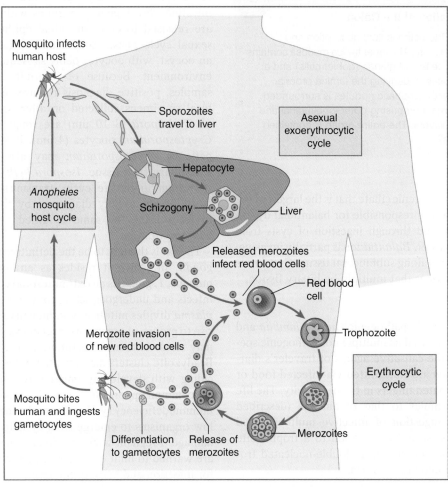

Figure 16-4. The *Plasmodium* (malaria) life cycle involves human and mosquito (*Anopheles*) hosts. Human hosts are infected by mosquitoes during a blood meal, following which sporozoites infect hepatocytes. Maturing schizonts split open hepatocytes, causing release of merozoites, which go on to infect red blood cells. Red blood cells from ring-stage trophozoites mature into schizonts. Alternatively, trophozoites can mature into gametocytes, which may be ingested by mosquitoes during feeding. Multiplication in the mosquito host leads to a series of morphologic changes resulting in sporozoite formation for infection of humans, thus completing the life cycle.

and slight hepatosplenomegaly. Late (chronic) infection occurs many years after initial infection and can lead to progressive encephalitis, dementia, and myocarditis. Organs of the digestive tract may also be affected. Related protozoan hemoflagellates, *Trypanosoma brucei gambiense* and *Trypanosoma brucei rhodesiense*, are causative agents of sleeping sickness, common to West and East Africa. These agents are spread by the tsetse fly (*Glossina*); CNS involvement and dementia are seen in late-stage disease.

Leishmania

Leishmaniasis is caused by the obligate intracellular protozoan *Leishmania*, of which more than 20 different species are known to induce human disease. The organism is transmitted by sandflies. *Leishmania donovani*, *L. mexicana*, *L. tropica*, and *L. major* are major causes of disease. Each species is somewhat distinct and can be identified serologically. Upon a blood meal, promastigotes are transferred to the human host; promastigotes develop inside macrophages. There are two main forms of disease: cutaneous leishmaniasis and visceral leishmaniasis (also called *kala-azar*). Cutaneous leishmaniasis is characterized by skin lesions, whereas kala-azar presents as a more severe disease with hepatosplenomegaly and loss of white blood cells.

Trichomonas

Trichomonas vaginalis is a flagellated protozoan that replicates in the genitourinary tract of males and females. It is not carried by vector arthropods but rather is spread by sexual intercourse and contact. It does not form cysts but has a characteristic trophozoite form with a large nucleus, four anterior flagella, and an undulating membrane. Treatment with a single dose of metronidazole is sufficient to clear infection.

Acanthamoeba and Naegleria

Acanthamoeba spp. and *Naegleria fowleri* are free-living, freshwater amebae that cause a potentially lethal CNS disease referred to as *primary amebic meningoencephalitis*. The *Acanthamoeba* spp. are similar to *Entamoeba*; however, infection leads to a more chronic form of disease, identified by pathologic manifestations of chronic granulomatous encephalitis. *N. fowleri* trophozoites infect humans by entering the olfactory neuroepithelium, eventually reaching the brain to cause disease with CNS involvement. *Balamuthia mandrillaris* is a related organism that may cause disease in immunocompromised individuals.

KEY POINTS ABOUT PROTOZOA

- Intestinal protozoa are causative agents of gastrointestinal disorders.
- Extraintestinal protozoa disseminate within the host to alternate sites, leading to varied pathology and tissue damage.

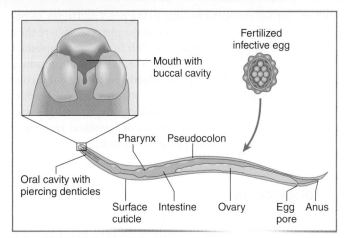

Figure 16-5. Structure of the roundworm.

NEMATODES

The nematodes are commonly referred to as *roundworms*. They range in size from a millimeter in length to over a meter. The male is typically smaller than the female. One end of the parasite has an attachment appendage, such as oral hooks or plates within the oral cavity (buccal capsule). The body is covered by a surface cuticle (Fig. 16-5).

Pinworms and Whipworms

Enterobiasis (pinworms, *Enterobius vermicularis*) and trichuriasis (whipworms, *Trichuris trichiura*) can be acquired through environmental contact with soil or food contaminated with eggs. The infective eggs are ingested, and the larvae hatch within the small intestine. The adult worms live in the cecum and ascending colon. For pinworm, the adult female worm is slightly larger (8 to 12 cm) than the male (3 to 5 cm); the adult whipworms are roughly 4 cm in length. Clinically, infection is manifested as abdominal pain and diarrhea that result from reactivity to egg deposition; females can shed between 3000 and 20,000 eggs per day.

Ascaris

Ascariasis is caused by *Ascaris lumbricoides*, the largest roundworm able to parasitize the human intestine (up to 30 to 35 cm in length). The human infection cycle begins with ingestion of infective eggs, after which hatched larvae first invade the mucosa of the intestine and are then carried through lymphatics to the portal vein and finally to the lungs. The larvae mature in the lungs, penetrate the alveolar walls, and ascend the bronchial tree to the throat, where they are swallowed. Adult worms mature in the small intestine, where the female reportedly can lay up to 200,000 eggs per day. The eggs are passed from the host with the feces. *Ascaris* infection is the most common nematode infection; treatment with albendazole or mebendazole is successful.

Hookworm

The two roundworm nematode species, *Ancylostoma duodenale* and *Necator americanus*, are the causative agents of hookworm. Filariform larvae physically penetrate the skin and are carried through the veins to the heart and lungs, eventually ascending the bronchial tree to the pharynx, where they are swallowed. Maturation to the adult form is accomplished in the small intestine, where adult worms mate and produce eggs. An early clinical feature of infection is that of cutaneous larva migrans, a form of acquired dermatosis that occurs during skin penetration and migration of organisms to other tissue sites. Blood loss through destruction of intestinal tissue is a common clinical feature of chronic infection.

Strongyloides

Strongyloidiasis is caused by the intestinal nematodes *Strongyloides stercoralis* and *Strongyloides fuelleborni*. Infective filariform larvae penetrate soil-exposed skin. Larvae travel to lung tissue, enter airways, and are swallowed. Adult worms mature in the intestines, and females begin to lay eggs. Eggs hatch in the intestinal tract to produce rhabditiform larvae. Larvae may be released to the soil where free-living adults can mate and lay eggs. Alternatively, larvae may be available for autoinfection and continuation of the infective cycle. Although *Strongyloides* is an intestinal nematode, the larvae can spread to the respiratory tract, resulting in hemoptysis and pulmonary infiltrates. The organism is eosinophilic on Papanicolaou (Pap) stains and ranges from 400 to 500 μm in length in cytologic preparations. Its internal structure, a closed gullet, and V-shaped notched tail are characteristic for diagnostic purposes. Treatment is usually with albendazole or thiabendazole.

Trichinella

Trichinellosis (trichinosis) is caused by the nematode *Trichinella spiralis*, acquired by eating cyst-contaminated meat (typically uncooked pork). Released larvae enter the intestinal mucosa and develop into adult male and female worms. The cycle continues as larvae released from adult worms enter the lymphatics and mesenteric veins, travel to the heart and lungs, and then enter the arterial system, which distributes them throughout the body. The final stage is entry to striated muscle, where they penetrate fibers and encyst. Viable cysts can survive for decades within the host. Clinical symptoms include initial gastrointestinal disorders (nausea, diarrhea, abdominal pain) and chronic complications due to larval migration and encystations in muscle tissues. Diagnosis may be made by muscle biopsy or by screening for antibodies to pathogenic antigens.

> **PATHOLOGY**
> **Cutaneous Larva Migrans**
> Cutaneous larva migrans is a parasitic skin disease caused by filarial hookworm larvae that penetrate human skin and then migrate subcutaneously, leaving extremely itchy red lines that may be accompanied by blisters (also called *creeping eruption* or *ground itch*). The reaction is characterized by erythematous, serpiginous (having wavy, white-dotted lesions with indented margins), and pruritic (itchy) dermatitis.

Filariae

Filarial pathogenesis (filariasis) is due to the presence of filaria-stage organisms in the circulation and most often is characterized by inflammation of the lymphatic vessels. Three specific organisms are major causes of disease: *Wuchereria bancrofti* and *Brugia malayi* cause lymphatic filariasis, and *Onchocerca volvulus* causes river blindness (onchocerciasis). Other species also cause pathogenesis, including *Loa loa*. The clinical manifestations include extreme lymphedema and elephantiasis, in large part due to immune reactions within draining lymphatics that occlude vessels.

The life cycle of filariae begins with the arthropod vector (mosquito, midge, or blackfly) spreading infection during a blood meal. The larvae migrate to the appropriate site of the host's body, where they develop into adults. Females produce microfilarial offspring that circulate in the blood, skin, or eye (in the case of *O. volvulus*) and can migrate to subcutaneous tissue. It is critical to understand life-cycle stages and periodicity of the microfilariae when considering diagnosis. Blood stages can be identified by smear staining, but skin snips also are a useful tool for detection of microfilariae.

●●● CESTODES AND TREMATODES
Taenia and Echinococcus

The cestodes include *Taenia saginata* (the beef tapeworm) and *T. solium* (the pork tapeworm) and *Echinococcus granulosus* (the hydatid worm). Both *Taenia* spp. have similar life cycles and their eggs are morphologically indistinguishable. Pigs serve as the intermediate host for *T. solium*; humans are infected when they ingest an immature tapeworm (a cysticercus) in undercooked pork, leading to harboring of adult sexually active worms within the intestinal tract. However, if humans ingest *T. solium* eggs, hatching cysticerci lead to cysticercosis. In this case, larvae produced in the small intestine migrate to other tissues. Resulting cyst forms may lead to tissue-related disease, such as neurocysticercosis when brain tissue is affected.

E. granulosus is the cestode tapeworm agent responsible for causing cystic echinococcosis, a cystic disease that can have a strong hepatic obstructive component in human infection. The human is an intermediate host, with members of the canine species being the definitive host. Also known as "hydatid worm," related members of the genus are known to cause alveolar echinococcosis and polycystic echinococcosis. Eggs ingested by an intermediate host hatch into larvae that travel through the blood and form hydatid cysts in the host's tissues (liver, lung, or brain). In the definitive host, the larvae transform into full tapeworms, which pass infective eggs in the feces.

Schistosomes

Three main species of *Schistosoma* trematodes are the causative agents for schistosomiasis. They are *Schistosoma haematobium*, *S. japonicum*, and *S. mansoni*. The schistosomal life cycle is depicted in Figure 16-6. Eggs released environmentally with feces hatch, and resulting miracidia infect a freshwater snail

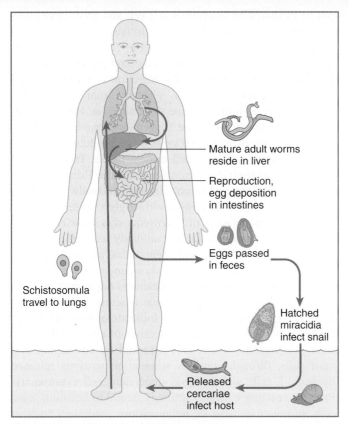

Figure 16-6. Life cycle of *Schistosoma*.

vector. Free-swimming cercariae released from the snail infect humans wading through infected waters, using an elastase to bore into the skin of the host. Cercariae will transform to schistosome form, travel to lung tissue, and eventually come to reside in the portal blood, where they develop into adult male and female worms. Specifically, *S. mansoni* and *S. japonicum* are frequently found in the superior mesenteric veins draining the large and small intestine, respectively, whereas *S. haematobium* often invades the bladder venous plexus. In all cases, organisms may reside in multiple tissue sites.

Damaging pathology ensues from immune responses to the eggs, with a resultant granulomatous response. A characteristic occurrence during infection is eosinophilia, which is thought to be induced by specific interleukins released by a skewed T-cell response. Over time, heavy fibrosis, calcification, and tissue scarring occur, often resembling cross-sectioning of a pipe stem, and lead to hepatic enlargement and severe hypertension. Clinical symptoms of hematuria and squamous cell carcinoma are common. Praziquantel and oxamniquine historically have been effective for treatment of disease-causing pathogens.

KEY POINTS ABOUT NEMATODES, CESTODES, AND TREMATODES

- The nematodes (roundworms) are generally large parasites that generate adult worms producing larvae or eggs that are damaging to the host.
- Cestodes and trematodes include the tapeworms and schistosomes, respectively.

- In all cases, adult worms give rise to eggs and/or cysts that cause physical damage, in part due to immunologic responsiveness to foreign antigens.

●●● LICE AND MITES

Humans are hosts to the head louse (*Pediculus humanus capitis*), the body louse (*Pediculus humanus humanus*), and the crab louse (*Phthirus pubis*). Lice attach to hair shafts and lay eggs (nits). Lice derive nutrients through tissue by blood-feeding and cause itching and irritation (Fig. 16-7). They are readily treatable with over-the-counter chemical applications.

Mites are related to ticks and spiders. There are many species, with commonalities in biting and bloodsucking behavior causing irritation and discomfort; many species are also known to elicit allergic reactions. Mites are spread by direct contact. Common mites that affect human hosts include *Sarcoptes scabiei* (scabies), *Eutrombicula alfreddugesi* (chiggers), and *Dermatophagoides pteronyssinus* and *Dermatophagoides farinae* (house dust mites). Treatment can be as simple as administration of oral antihistamines and application of hydrocortisone cream to relieve associated symptoms.

HISTOLOGY

Liver

The liver is a filter for toxic metabolites. The parenchyma is comprised of liver plates (hepatocytes) with endothelial cells establishing sinusoids for blood flow. The characteristic portal triad contains at least one profile each of a hepatic artery, a portal vein, and an interlobular bile duct.

●●● LABORATORY DIAGNOSIS OF PARASITIC INFECTIONS

Identification of parasites is critical for proper treatment and limitation of pathology within the host. Knowledge of the characteristics of clinically significant human parasites includes recognition of organism morphology at various stages of life cycles and an understanding of the relationship of morphology to disease state. In addition, the medical practitioner

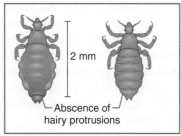 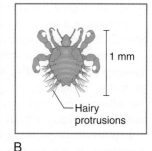

Figure 16-7. A, The common head and body louse (*Pediculus humanus*). **B,** The crab louse (*Phthirus pubis*).

must understand potential routes of infection and appreciate the value of appropriate clinical specimen collection to diagnosis of disease.

Many diagnoses begin with macroscopic examination of fecal material, with regard to consistency of stool specimen and presence of intestinal protozoa or visual inspection for eggs and larvae. Care should be taken to limit contamination from environmental sources and from patient urine. Adult worms are more easily detectable than ova and smaller organisms, which require microscopic examination and separation of stool into components for identification by staining techniques. The Wheatley trichrome technique is a rapid way to stain for intestinal protozoa, allowing identification of cysts. *Cryptosporidium* is more easily identified with a modified Kinyoun acid-fast stain. Pinworms lay eggs in the perianal region and may be recovered using a simple cellophane tape method. For blood-borne parasites, serology is the simplest means of detection. Although culture of organisms is usually possible, detection of host antibodies against different-stage proteins reveals information on tissue migration and developmental stage and may also indicate relative numbers and activity of infection. Standard enzyme-linked immunosorbent assays are a quick way to determine current or past exposure and are available for most common pathogens. Finally, blood smears are a rapid way to determine presence of schizogony for organisms such as *Plasmodium, Babesia, Leishmania,* and *Trypanosoma*. Fixed peripheral blood is treated with Giemsa stain to allow visualization of intracellular parasites or of microfilariae.

IMMUNE RESPONSE TO PARASITIC INFECTIONS

Host immune responses to parasitic infections are dictated by organism size, location within the host, and parasite complexity. Natural barriers provide the first line of protection, and portals of mucosal entry are lined with degradative enzymes and immunoglobulins (Ig) (e.g., IgA). For example, the immune response to intestinal protozoa is primarily humoral and inflammatory. Protozoa such as *Cryptosporidium* and the *Entamoeba* spp. stimulate an immune response within the gut-associated lymphoid tissue. Initial reactivity begins with a nidus of tissue damage and ensuing activation of enterocytes, characterized by release of cytokines and chemokines. Neutrophils are a first line of defense once organisms have invaded tissue, phagocytosing when they can, or releasing factors (cationic defensins, lactoferrin, and myeloperoxidases) that have direct cytotoxic effect. Chemotactic factors, including breakdown products of complement activation, attract dendritic cells and macrophages (antigen-presenting cells), which initiate recognition by lymphocytes. In most cases, lymphocytes induce production of proinflammatory responses while triggering B cells to produce both IgA and IgG antibody subclasses.

Extraintestinal protozoa represent further requirements for complexity in immune responses. The malaria parasite changes morphology during its life cycle in humans, which requires different immune interventions. Although blood-stage protection can be mediated by antibodies, a T-cell and productive interferon response against blood-stage antigens may be required for complete protection against disease. *Toxoplasma* is an organism that easily crosses encephalic and placental barriers. The limited acute phase of this infection is associated with strong interferon responses elicited from T lymphocytes, effectively limiting tissue dissemination. However, this specific lymphocyte response triggers an organism life-cycle change resulting in manifestation as an intracellular bradyzoite that may persist in brain and muscle tissue. Trypanosomes bring about an even more complex response. Primary resolution of trypanosomal infection is through antibody recognition of surface antigens; variations in surface glycoproteins through molecular rearrangement of DNA sequences allow replicating organisms to escape specific host antibody recognition.

Nematodes, cestodes, and trematodes undergo physical changes as they adapt to life in the host, and they are therefore subject to differing immune mechanisms depending on form (larval, adult, or egg stage). Tissue reaction to *Ascaris* and *Trichinella* consists of an intense infiltrate of polymorphonuclear leukocytes, with a predominance of eosinophils. Likewise, schistosomes are readily attacked by eosinophils and mast cells, through complex stimulation patterns released by both T cells and antibody-dependent cell cytotoxicity. Proteins reacting with IgE antibody bound to intestinal mast cells stimulate release of inflammatory mediators such as histamine, proteases, leukotrienes, prostaglandins, and serotonin. However, schistosomal eggs trigger responses from macrophages and T cells, leading to development of a granulomatous physical encapsulation of slowly released, nondigestible antigenic materials.

KEY POINTS ABOUT PARASITOLOGY

- It is extremely important to understand the life cycle and have strong knowledge of the developmental stages of parasites for positive diagnostic identification to occur.
- Immunologic mechanisms to combat parasite pathogens are varied depending on the size and number of organisms and relative location of directed responses.

KEY POINTS

- Parasitic organisms may be facultative or require obligate interactions with the host. Knowledge of the life cycle and developmental stages of parasites are critical for diagnostic identification. Intermediate hosts (or vectors) serve only as temporary reservoirs, allowing physical metamorphosis to reach a human infective stage.
- Intestinal protozoa are causative agents of gastrointestinal disorders, usually transmitted by ingestion of contaminated food or water. Organisms subsequently disseminate from the gut to alternate sites, leading to varied pathology and tissue damage. *Entamoeba histolytica* and *Cryptosporidium parvum* are two examples of this organism class.

- Extraintestinal protozoa spread by vectors include the *Plasmodium* species, the causative agents of malaria. *Trypanosoma cruzi* (Chagas disease) and *Leishmania* spp. are transmitted to humans by blood-sucking insects, after completion of life-cycle stages in the arthropod host.
- The nematodes (roundworms) are large parasites. The adult worms produce larvae or eggs. Ascariasis, caused by *Ascaris lumbricoides*, is the largest roundworm able to parasitize the human intestine. Strongyloides represents an infection that can affect both intestinal and pulmonary systems. *Onchocerca volvulus* causes river blindness (onchocerciasis); *Brugia* is the agent responsible for filariasis.
- Cestodes and trematodes include the tapeworms and schistosomes, respectively. An example of a cestode with clinically relevant importance is *Taenia*, which causes cysticercosis. *Schistosoma* spp. are the causative agents of schistosomiasis. In many cases, ensuing pathology following infection with these organisms is due to immunologic responsiveness to foreign egg antigens deposited in host tissue.
- Many diverse immunologic mechanisms to combat parasite pathogens have evolved, dependent upon the size of the invading organism and the relative physical tissue location where directed responses occur.

Self-assessment questions can be accessed at www.StudentConsult.com.

Case Studies

CASE STUDY 1

A 14-year-old male middle school student was admitted to the emergency department with centered abdominal pain that progressed to the right lower quadrant. Additional symptoms included vomiting and slight diarrhea. Physical examination revealed tenderness with guarding (muscle tensing) upon touch. Blood tests showed a slightly elevated white blood cell (WBC) count of 16,300/mm^3 (normal, <12,000/mm^3). A diagnosis of appendicitis (inflammation of the appendix) was made, and an appendectomy was immediately performed.

1. What is the function of the appendix?

2. What is the common cause of appendicitis?

3. What are complications if appendicitis is left untreated?

CASE STUDY 2

An 11-month-old male presents with rapid onset of high fever and elevated WBC count due to an upper respiratory infection with *Haemophilus influenzae*. The child has a history of recurrent bacterial infections with encapsulated organisms. Upon clinical examination, the spleen was not palpable. A diagnosis of congenital asplenia was made.

1. How does the lack of a spleen affect immune cell function, and what implications does this have for immune responses to infective agents?

2. Why does the lack of a spleen represent more of a clinical problem in children than in adults?

3. What is the appropriate medical care and response for congenital asplenia?

CASE STUDY 3

A 12-year-old female gymnast developed pain in her upper right thigh, which increased progressively during stretching before floor exercise competition. During competition, she landed incorrectly and broke her leg. Radiographs reveal a fractured femur; however, spongy-appearing lesions are also seen throughout the long bones of her leg. Laboratory analysis shows mild anemia and neutropenia with a decreased WBC count of 3200/mm^3 (normal, 5000/mm^3) and highly elevated immunoglobulin G levels at 4250 mg/dL (normal, 700 to 1000 mg/dL). The diagnosis is multiple myeloma, and she is started on a chemotherapeutic regimen.

1. What is the immunologic basis for multiple myeloma?

2. What is the pathologic basis for the spongy appearance of bone tissue?

3. In the face of high antibody levels, why are patients considered more susceptible to bacterial infections?

CASE STUDY 4

An 8-month-old male presents to the pediatrician with recurrent ear and sinus infections. Physical examination reveals distinctive downward-slanting eyes; low-set, rotated ears; and a small, round mouth. Radiographs and computed tomography (CT) scans demonstrate absence of a thymic shadow. Laboratory analysis of WBCs reveals chromosomal defects, and the child is given the diagnosis of DiGeorge syndrome.

1. What is the pathologic basis of DiGeorge syndrome?

2. What is the effect of thymus absence on maturation of T cells?

3. What is the prognosis for patients with DiGeorge syndrome?

CASE STUDY 5

A 53-year-old African American male with a history of emphysema underwent a single lung organ transplant. He has been on a cyclosporine and corticosteroid–based regimen to limit the possibility of graft rejection. His postoperative course had been unremarkable until the recent development of a cough with production of heavy yellow sputum.

He reports fatigue but does not have night sweats, fever, or chills. Rhonchi (leathery wheezing) can be heard within the transplanted lung; a radiograph reveals extensive interstitial infiltrates. A bronchoscopic biopsy specimen does not show evidence of rejection, nor is there evidence of granulomas (a chronic inflamed tissue mass). However, viral inclusions are apparent, and the eventual diagnosis is parainfluenza as the cause of the upper respiratory infection and bronchitis.

1. How do immunosuppressive drugs contribute to the pathology seen in this case?

2. What is the molecular basis for graft rejection in patients receiving donor tissue?

3. What other viral associations are commonly found in patients who have undergone organ transplantation?

CASE STUDY 6

A 14-month-old male is brought to the pediatrician because of a deep and painful perirectal abscess. His WBC count is greater than $45,000/mm^3$ (normal, less than $10,000/mm^3$), and his spleen is enlarged (splenomegaly). His parents and siblings are healthy. Lymphocyte function, antibody formation, nitroblue tetrazolium test (NBT) results, and complement function are normal. Further examination of WBC populations reveals a deficiency in chemotactic activity, and a diagnosis of leukocyte adhesion deficiency (LAD) is made.

1. What is the molecular basis for LAD?

2. What is the function of the NBT test?

3. What is the prognosis for LAD?

CASE STUDY 7

A 4-month-old female is referred to City Hospital after 3 weeks of prolonged fever episodes and sores in the upper mouth and on the tongue. Immunophenotypic analysis of blood cells identifies a marked eosinophilia and a deficiency in the mature T-lymphocyte and B-lymphocyte population. In addition, serum antibody levels are low for all immunoglobulin isotypes. Proliferative responses to mitogens (concanavalin A and lipopolysaccharide) are markedly reduced. Based on clinical, immunologic, and genetic findings, Omenn syndrome is diagnosed.

1. What is the molecular basis for Omenn syndrome?

2. Why do some patients with Omenn syndrome demonstrate eosinophilia?

3. What is the prognosis for Omenn syndrome?

CASE STUDY 8

A 34-year-old male reported sudden onset of weakness in the extremities, coincident with increased work-related stress. The patient demonstrated a slightly elevated WBC count but no increase in banded patterns (polymorphonuclear cells) and no fever. Culture tests did not identify a causative infectious agent. Symptoms resolved with rest and a return to normal routine. Now, 5 years later, after an especially stressful month of work, the symptoms returned with noticeable loss of upper arm strength and awkwardness in gait. His WBC count is elevated, with a right shift indicative of increased lymphocytic responses. The increased neurologic dysfunction warrants magnetic resonance imaging, upon which evidence of plaque formation in the central nervous system is found. Early-stage multiple sclerosis is diagnosed.

1. What is the immunologic basis for multiple sclerosis?

2. What is meant by the diagnostic indication of plaques?

3. What factors contribute to development of multiple sclerosis?

CASE STUDY 9

A 19-year-old college student reports shortness of breath, swollen gums, and persistent fever. He also indicates that he bruises easily, at times with no apparent cause, and reports abnormally long times for cuts to heal. Close examination of skin tissue reveals small petechial hemorrhages under the skin. Blood tests lead to the diagnosis of acute myelogenous leukemia (AML). He immediately undergoes extensive and aggressive radiation therapy and chemotherapy. He receives a bone marrow transplant from a sibling and is given a short course of immunosuppressive therapeutics.

1. What is AML?

2. What are possible side effects of bone marrow transplantation?

3. What are the ABO blood group antigens?

4. Why was the patient given a course of immunosuppressive therapeutics?

CASE STUDY 10

A 16-year-old Chinese female awoke with slight fever. She took two aspirin and went to school; however, by mid-morning she had a diffuse macular erythroderma (rash) on her extremities. A noninvasive examination by the school nurse indicated no abnormal use of drugs or alcohol; however, it was readily apparent that she was disoriented and in need of emergency care. En route to the hospital, she reported severe myalgia (muscle pain) and lost consciousness as a result of hypotension. Laboratory tests indicated elevated creatine phosphokinase and alanine aminotransferase levels as well as pyuria (pus in the urine). *Staphylococcus aureus* was isolated from blood cultures. Toxic shock syndrome (TSS) is diagnosed.

1. What is toxic shock syndrome?
2. What is the immunologic basis for toxic shock?
3. What was the basis of her physical symptoms?

CASE STUDY 11

A 76-year-old retired male is admitted to the emergency department for treatment of cellulitis on the top of his left foot and the appearance of a blistering hematoma (5 × 5 cm) with radiating erythema on his left heel. Complete blood count studies reveal a left shift, with elevated WBC count of 23,800/mm^3 (normal, <12,000/mm^3). The radiograph of the ankle and foot shows extensive gas in the soft tissues. A biopsy specimen taken from the wound area reveals few neutrophils in spite of the high WBC count. On gram-positive staining, boxcar-shaped rods are identified that do not grow on aerobic cultures.

1. Which organism is responsible for the disease?
2. Which factors contribute to the pathology?
3. What is the course of action for treatment?

CASE STUDY 12

A 49-year-old female caseworker for child protective services reports persistent cough, night sweats, loss of appetite and weight, and occasional blood-streaked sputum. The patient has a 20-year history of smoking (one pack per day). Cardiopulmonary examination is unremarkable. Her WBC count is slightly elevated at 16,700/mm^3 (normal, 4500 to 10,500/mm^3), but blood chemistry levels are within normal range. Chest radiograph reveals lung infiltrate, and a CT scan confirms thick-walled cavitation in the upper right lobe. Acid-fast organisms grow from cultured sputum and bronchoalveolar lavage fluid and are identified as *Mycobacterium tuberculosis*.

1. What is the immunologic basis for the lung cavitation?
2. Which factors in the patient's history contribute to disease-related pathology?
3. What therapeutic intervention is prescribed for this individual?
4. Why do these organisms stain acid-fast?

CASE STUDY 13

An otherwise healthy 29-year-old white female presents with a 3-day history of headache, fever, stiff neck, and photophobia (painful sensitivity to light). She shows general malaise and lethargy. The patient also indicates nausea with two incidents of vomiting and watery diarrhea persisting for almost a week. She has no history of sexually transmitted diseases and is HIV negative. A lumbar puncture is performed, revealing lymphocytic pleocytosis (increased cellularity) in the cerebrospinal fluid. Polymerase chain reaction and antibody reactivity tests reveal that she has aseptic meningitis from reactivated herpes simplex virus 2 (HSV-2) infection.

1. What is the likely history of this patient?
2. What effects does HSV-2 have on the brain and central nervous system?
3. Which events led to reactivation of infection?

CASE STUDY 14

A 28-year-old Hispanic male presents to City Hospital with crampy right-lower-quadrant abdominal pain, recent excessive weight loss, fever, fatigue, and nonbloody diarrhea. The patient denies intravenous drug use or homosexual activity, but he admits to recently serving 2 years in the county jail. He has multiple tattoos and a history of stab wounds. Laboratory tests indicate the patient is HIV positive, and he admits to being noncompliant with highly active antiretroviral therapy (HAART). He has a viral load of more than 300,000 copies/mL and a CD4 count of approximately 120 cells/mm^3. His laboratory test results are negative for purified protein derivative (PPD) at the time of testing. A CT scan of the abdomen shows thickening of the cecum and ascending colon but no lymphadenopathy.

1. Which organism(s) might be responsible for the disease?

2. What is the course of action for treatment?

3. What is HAART therapy?

CASE STUDY 15

A 40-day-old female neonate born 6 weeks premature underwent corrective surgery to remove an oral cyst. The surgery was successful; however, 4 days later the infant has fever, tachycardia (rapid heart rate), tachypnea (increased respiration rate), and thrombocytopenia (decreased blood platelets). The area around a percutaneously inserted central line appears swollen and red with obvious phlebitis (inflammation of vein). Blood cultures taken from the central line are positive for *Candida albicans*; the child is immediately started on antifungal therapeutics, and improvement in her condition is noticed shortly thereafter.

1. What are predisposing risk factors for candidiasis in the neonate?

2. What is the likely source of contamination for this child?

3. What are complications if the fungal infection is not treated effectively and immediately?

CASE STUDY 16

A 35-year-old male woodworker living in the Blue Ridge Mountains of Tennessee comes in with right-upper-quadrant pain, slight nausea, and a 2-week history of diarrhea. Visual examination shows calloused hands, the presence of tattoos on his right forearm, and a slight yellowish tint to his sclera. He is a Honduran immigrant who has been in the United States for 6 years. He admits to alcohol consumption of at least two beers per day and indicates he spent 8 months in prison in 2002. His laboratory test results are negative for PPD and HIV. Blood tests show an elevated leukocyte count of 7600/mm^3, with 54% neutrophils (within normal range) and 12% eosinophils (highly elevated).

1. Which organ systems are affected, and what could be the possible cause of the disorder?

2. What complications are associated with this type of infection?

3. What is the recommended course of treatment?

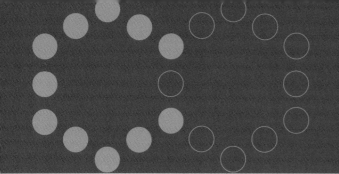

Case Study Answers

CASE STUDY 1

1. The appendix (as well as tonsils and Peyer's patches) is a representative lymphoid tissue found in and around mucosal epithelia (referred to as mucosa-associated lymphoid tissue [MALT]). Within the appendix, specialized resident intraepithelial lymphocytes with potent cytolytic and immunoregulatory capacities monitor mucosal tissue to help defend against pathogenic infection.

2. Obstruction of the appendix lumen is the primary cause of appendicitis, leading to tissue distention from excessive accumulated intraluminal fluid. The most common cause of luminal obstruction is lymphoid follicle hyperplasia, at times reflective of high bacterial burden and overaggressive immune response. Ineffective lymphatic drainage allows bacterial invasion within the appendix wall, with high polymorphonuclear cell reactivity and resultant inflammation.

3. Standard therapy involves removal of the inflamed appendix through a procedure aptly named *appendectomy*; serious complications may result if the appendix ruptures. A ruptured appendix can lead to peritonitis from leakage of trapped bacteria into the abdomen. Often, abscess formation will occur, comprising a swollen mass filled with fluid and bacteria and pus (dead and dying neutrophils trying to control spread of infection). Peritoneal bacteria often release endotoxins, such as lipopolysaccharides, contributing to septic events and the possibility of secondary organ failure.

CASE STUDY 2

1. The deficiency is a compromise of both the reticuloendothelial system and adaptive immunity due to the congenital absence of a spleen. The major shortcoming is a loss of mononuclear phagocytes to clear pathogenic organisms from the blood. Secondarily, there is a loss of B- and T-cell mass, which ultimately may be largely compensated at the level of systemic lymphoid tissue. The individual is predisposed to morbidity and mortality from encapsulated organisms such as *Streptococcus pneumoniae*, *Haemophilus influenzae*, and *Neisseria meningitides* (meningococcus), which require antibody response for protection.

2. Loss of a spleen would be more detrimental to a child than to an adult primarily because of a previously established immune response (B cells and their ability to produce specific antibodies) to bacterial antigens in the adult. In the adult, preexisting memory B cells surviving in other tissues (e.g., lymph nodes, gut-associated lymphoid tissue [GALT], MALT, and bronchus-associated lymphoid tissue [BALT]) may be activated, although the overall response in these adults typically is diminished. In the child, there is less likely to be a preexisting memory population. The lack of splenic lymphoid tissue to process antigen greatly decreases the opportunity for B-cell activation and further production of plasma cells and memory B cells. Additional concerns include lowered specific immunoglobulin production and decreased complement fixation associated with the classical component pathway.

3. Medical care involves antibiotic prophylaxis, immunizations, aggressive management of the suspected infection, and parent education. Antibiotic prophylaxis is initiated immediately upon the diagnosis because patients are considered immunocompromised. Patients should receive standard immunizations, especially conjugated *H. influenzae* type b and pneumococcal vaccines. Asplenic patients are at an increased risk for sepsis, especially from gram-positive organisms. Parents must be educated to remain vigilant to detect symptoms of infection so that early detection will allow aggressive management.

CASE STUDY 3

1. Multiple myeloma (also called *plasma cell myeloma*) is a progressive cancer of the plasma cell, a mature B cell that functionally produces immunoglobulins. Multiple myeloma is characterized by excessive numbers of abnormal clonal plasma cells in the bone marrow and overproduction of intact monoclonal immunoglobulin (of any isotype) or Bence-Jones protein (circulating monoclonal κ and λ light chains in the absence of heavy chains). Because the myeloma cells are identical, they all produce the same immunoglobulin protein, called *monoclonal (M) protein*, in large quantities.

2. Transformation of a B cell to a malignant plasma cell involves processes that generate genetic abnormalities, sometimes including translocation of chromosomes near the heavy chain locus. The outcome is that malignant cells are allowed

to divide unchecked. Myeloma cells that invade bone can establish masses, or plasmacytomas, that produce growth factors and vascular endothelial growth factor to promote angiogenesis. Eventually, multiple small lesions form and spread into large bone cavities throughout the body, giving rise to anemia and reduced polymorphonuclear cell production owing to displacement of functional hematopoietic stem cells.

3. Although high levels of circulating antibody exist, the antibodies produced are monoclonal in nature. The hypergammaglobulinemia comes at the relative expense of normal immunoglobulin production. In essence, the patient becomes functionally antibody-deficient in spite of high numbers of nonspecific antibodies in circulation. This, combined with anemic and neutropenic conditions, raises susceptibility to pyogenic bacteria requiring specific antibody response for clearance of infectious organisms.

CASE STUDY 4

1. DiGeorge syndrome is a rare form of primary immunodeficiency, the basis for which is congenital absence or aplasia of the thymus due to defects during embryogenesis stemming from a large deletion from chromosome 22.

2. T lymphocytes released from the bone marrow migrate to the thymus, where further development and education into functional phenotypes occur. Thymic education includes removal of self-reactive T lymphocytes and commitment of remaining cells to helper (CD4+) or cytotoxic (CD8+) phenotypes. In the absence of a thymus, maturation of T lymphocytes does not occur, and CD3+, CD4+, and CD8+ populations are particularly low.

3. The prognosis depends on the severity of T-lymphocyte compromise. Opportunistic infections occurring because of severe immunodeficiency are the second most common cause of death in patients with Di George syndrome. Although B lymphocytes are present and serum immunoglobulin G levels are typically normal, the functionality of the B-cell response is limited owing to lack of help from antigen-specific T helper (T_H) cells. Common infections manifest as ear and sinus infections, pneumonia, infectious diarrhea, and severe thrush (Candida). Antibiotic treatment assists in the short term, but long-term therapeutic intervention dictates strengthening the immune system by bone marrow transplantation and possibly thymus transplantation to recover maturation of the T-cell response.

CASE STUDY 5

1. It is common for individuals to be maintained on immunosuppressive agents for long periods after major organ transplantation. The symptoms described in this patient overlap with early stages of graft rejection; however, in this case the cause was viral. Immunosuppression is the critical postoperative factor that predisposes to infection in transplant recipients, as agents dampen immune function; patients typically remain on low levels of therapeutics for the rest of their lives. Dampening of T_H and T cytotoxic lymphocytic responses specifically limits the ability to fight against viral infection, thus contributing to the disease state in this person.

2. The human leukocyte antigens (HLAs) are a complex set of molecules expressed on the surface of cells. These molecules, also referred to as *major histocompatibility complex molecules* (MHCs), are products of polymorphic genes that are the principal determinants of whether an organ is deemed as self or nonself. These molecules present antigen to T lymphocytes as a method of regulating immune response. In general, class I MHC molecules present peptides derived from endogenous pathways to cytotoxic lymphocytes, whereas class II MHC molecules present exogenous antigens to T_H lymphocytes. Donor tissue must match closely in HLA alleles for acceptance by the recipient.

3. In addition to parainfluenza virus, other community-based respiratory viruses are commonly found in immunosuppressed transplant patients, including influenza, respiratory syncytial virus, herpesvirus, varicella virus, and adenovirus. Furthermore, cytomegalovirus (CMV) is quite common, most likely related to the patient's prior CMV status (upwards of 80% of the adult population have been exposed to CMV).

CASE STUDY 6

1. LAD is an autosomal recessive disorder in which there is a failure to express surface integrins on myeloid and lymphoid cells. Migration of activated polymorphonuclear cells and lymphocytes to sites of inflammation requires expression of selectins on endothelial cells and concurrent binding by ligands on rolling leukocytes. Directed diapedesis ensues through integrin interactions (e.g., via CD11/CD18 surface β_2-integrins). Lack of CD18 results in inefficient entry of circulating leukocytes to sites of inflammation. Lack of leukocyte function antigen-1 would result in similar immune deficiency. In addition, CD18 serves as a complement receptor on myeloid cells (for C3b), assisting with phagocytosis and efficient microbicidal activity in neutrophil, monocyte, and macrophage populations.

2. The NBT test allows examination of bactericidal activity of polymorphonuclear cells. NBT is an electron acceptor that is reduced in the presence of free O_2 radicals to form a blue compound, formazan. NBT is added to activated white blood cells. Neutrophils produce reactive oxygen species (ROS) when stimulated, a function that assists in bactericidal activity. The ROS produced also turns NBT from clear to deep blue. In this patient, there is no deficiency in production of ROS; rather, additional assays of random migration and chemotaxis were necessary to define phenotypic deficiencies.

3. Most patients with LAD have a poor outcome, with few infants surviving past 1 year of age. Bacterial infections and severe viral infections are responsible for most deaths. The most common bacterial agents are the *Staphylococcus* species and enteric gram-negative bacteria. Infection with fungal agents (such as *Candida albicans*) is also common. In many cases, bone marrow and other stem cell transplantation procedures have proved effective at prolonging survival.

CASE STUDY 7

1. Omenn syndrome is an autosomal recessive form of severe combined immunodeficiency (SCID) characterized by susceptibility to fungal, bacterial, and viral infections. The molecular defect is linked to defects in either of the recombinase-activating genes (RAG), *RAG-1* or *RAG-2*, which control generation of antigen-binding diversity by assembly of the genes encoding variable (V), diversity (D), and joining (J) domains for both the T-cell receptor (TCR) and the B-cell receptor (antibody). The inability to productively rearrange exons required for T- and B-cell receptors leads to poorly functional T cells and absent B cells.

2. The remaining *RAG* gene remains functional, albeit deficient in complete maturation of lymphocytes without the other gene. It is hypothesized that a subpopulation of T_H lymphocytes of the T_H2 variety is expanded, possibly as a result of increased exposure to inadequately cleared antigens and resulting expansion of a minor oligoclonal population of activated and antigen-stimulated cells. These T_H2 cells produce elevated levels of interleukin-4 (IL-4) and -5 (IL-5), cytokines that mediate eosinophilia. Of interest, high production of nonspecific IgE (hyperimmunoglobulinemia E) may also occur in these patients. Likewise, the limited repertoire of productive lymphocytes is likely to show decreased responses to specific antigens and variable responses to nonspecific mitogens.

3. Mortality and morbidity levels are increased in all SCID patients, regardless of the specific molecular defect. Life-threatening viral, bacterial, and fungal infections are prominent in Omenn syndrome patients, as in other types of SCID. A higher incidence of malignancies occurs, most likely due to loss of regulating cytotoxic lymphocytes. Prophylactic antibiotics are warranted. Bone marrow transplantation is usually successful to reconstitute immune surveillance; however, care should be taken to avoid the common complication of chronic graft-versus-host disease (GVHD).

CASE STUDY 8

1. Multiple sclerosis is a disease of the central nervous system that affects the nerves of the brain and spinal cord. Myelin, produced in the brain and spinal cord by oligodendrocytes, forms a protective coat around axons, allowing nondisruption of electric impulses away from nerve cells. It is thought that immunoreactive T lymphocytes contribute to destruction of the myelin sheath, recognizing self-tissue as foreign. This in turn assists in the production of autoantibodies from self-reactive B lymphocytes. Ultimately, this leads to disease manifestation.

2. The process involves destruction of myelin (demyelination), producing multiple patches of hard, scarred tissue called plaques; "sclerosis" (Greek) means hard. Sodium ion channels, which boost electrical charges between nerves, are clustered at the nodes of Ranvier. These channels are disrupted, leading to interruption of signal transmission between cells within the nervous system when sodium ions are not able to boost impulses.

3. It is not clear why some autoimmune events occur in everyone, yet not everyone develops autoimmune disease. Genetic factors certainly play a role in development of MS, with risk factors higher in first-degree relatives and identical twins. However, no single gene has been identified to trigger autoimmune conditions. Rather, a combination of genetic factors most likely renders individuals more susceptible to disease manifestation. It is also likely that environmental factors can trigger the autoimmune response, be it chemical or infectious in nature. Although certain infections are suspected to trigger autoimmune MS (e.g., herpesviruses), no single agent to date has been identified. Although they are not causative, both trauma and stress can exacerbate disease symptoms.

CASE STUDY 9

1. AML is a type of blood cancer in which myeloid cells become cancerous, usually due to DNA damage within developing cells of the bone. The resultant chromosomal abnormalities disrupt genes encoding for transcription, which are required for stem cell differentiation into specific blood phenotypes.

2. Bone marrow transplantation (also referred to as *stem cell transplantation*) is an excellent clinical tool to reconstitute normal immune and hematopoietic function in patients undergoing cancer chemotherapy. The major complications include increased susceptibility to infection from opportunistic organisms or latent viral entities due to lack of mature lymphocytes specific to ward off infecting agents. As new cells are generated in the host, the ability to fight infection returns. Another potential adverse effect can occur when the source of bone marrow used for transplantation is allogeneic (not syngeneic), leading to GVHD over time. The basis of GVHD is that mature donor lymphocytes transferred to the recipient may show reactivity to mismatched antigens (major histocompatibility antigens or ABO groups) residing on recipient tissue. In all cases, laboratory tissue typing and mixed-lymphocyte reactions (culturing donor and recipient cells together) would be performed to ascertain reactivity between new and existing cell populations.

3. The ABO blood groups are carbohydrate moieties present on erythrocytes that represent important antigens to be accounted for to ensure safe blood transfusions. Individuals naturally develop antibodies (called *isoantibodies*) specific for ABO antigens that they do not express. If the individual receives a transfusion of blood containing incompatible ABO antigens, isoantibodies will cause agglutination of the donor cells, called *isohemagglutination*.

4. The patient had undergone ablation of precursor stem cells, thus limiting his reactivity to the foreign donor cells. However, persistent peripheral cells within the host may have survived with enough reactivity to attack new donor cells; immunosuppressive agents were administered to allow the new donor stem cells to inhabit the bone marrow and limit destruction before successful reconstitution of host immune function.

CASE STUDY 10

1. TSS is a severe illness associated with *Staphylococcus aureus* infection, characterized by fever, nausea, diffuse erythema, and shock. It typically occurs in menstruating females using tampons, primarily when saturated tampons are not changed often, allowing unchecked growth of microorganisms. However, TSS is not limited to women and may occur in any individual exposed to large numbers of *S. aureus*.

2. The superantigens staphylococcal enterotoxin and toxin-1 are responsible for toxic shock syndrome. Superantigens are defined by their ability to stimulate a large fraction of T cells via interaction with the T-cell receptor (TCR), usually through the variable beta (Vβ) domain. In this manner, the superantigen circumvents normal processing and directly links the TCR with the MHC class II molecule, causing cells to differentiate into effector cells that release high levels of cytokines. Because the number of T cells that share Vβ domains is high (up to 10% of all T cells), large numbers of T cells (regardless of antigen specificity) may be activated by superantigens, causing massive release of cytokines and subsequent symptoms of shock and host injury.

3. The overwhelming number of cytokines released from activated T cells caused a cascade of effects at both local and systemic levels. Cytokines released include IL-2, tumor necrosis factor-α (TNF-α), and IL-1β. IL-2 is a mitogenic stimulatory factor for T cells, thus furthering the cascade of T-cell activity. Both TNF-α and IL-1β cause activation of vascular endothelium and increased vascular permeability, thus contributing to hypotensive shock in the patient as fluid accumulation in vascular beds occurred. Local responses also included direct tissue damage, since TNF-α can be a necrotic factor at high concentrations. These same cytokines, along with IL-6, evoke systemic effects including fever production, mobilization of shock-related metabolites, and induction of acute-phase protein production. This was manifested by elevated enzyme levels, which were further indicative of destruction of renal and hepatic tissue.

CASE STUDY 11

1. The finding of boxcar-shaped, gram-positive rods that do not grow aerobically is consistent with *Clostridium perfringens*, frequently associated with gas gangrene (see Chapter 12). Spores of the organism persist in soil; introduction of organisms into wound areas followed by toxin production destroys skin and muscle tissue. The bacteria grow and manufacture "gas" as a byproduct, often detectible by radiographs or scans of the wounded area. *C. perfringens* is also the causative agent for more common disease-associated food poisoning due to gut-released enterotoxins.

2. *Clostridium* spp. produce virulence factors, including a lecithinase that is capable of lysing polymorphonuclear cells, thus the lack of cellularity in the tissue biopsy. The α-toxin lecithinase (a phospholipase C) plays a central role in development of gas gangrene pathogenesis while limiting the innate immune system response. Additional tissue-destructive factors released include a protease, hyaluronidase, collagenase, and general hemolysins.

3. If infection is not treated immediately and aggressively, mortality rates are greater than 40%. Aggressive penicillin antibiotic therapy is warranted with accompanying surgical debridement. If the infection does not respond to this therapy, then amputation above the affected area is warranted to limit spread through draining lymphatics. Because the organism is anaerobic, hyperbaric chamber treatment (O_2-rich atmosphere) is useful.

CASE STUDY 12

1. Organisms engulfed by macrophages stimulate responses leading to exudative (pneumonia-like) or granulomatous lesions. The granulomatous lesions are characterized by giant multinucleated cells that are surrounded by lymphocytes, forming the basis of a tubercle. Immunologic responses to the persisting organism trigger destructive pathology and necrotic events, leaving a caseous center to the granuloma. Eventual erosion of lesions into bronchial airways leads to spread of disease.

2. Risk factors such as smoking habits, alcohol consumption, family history of tuberculosis, and lifetime work history all contribute to potential for disease. A social worker would have greater exposure to mycobacteria from indigent populations with limited access to health care. In this patient, smoking for many years most likely impaired native lung macrophage function, contributing to the progression from infection to manifestation of disease. Other factors, such as age and ethnic origin, are also associated with an increased risk of infection progressing to disease.

3. Because of the slow growth doubling of mycobacteria (18 to 24 hours), therapeutic intervention is made over a lengthy period. The Centers for Disease Control and Prevention recommend multidrug treatment with isoniazid, rifampin,

pyrazinamide, and ethambutol or streptomycin for tuberculosis for compliant patients with fully susceptible organisms. At least one of these drugs are taken daily for 2 months followed by daily isoniazid and rifampin therapy for an additional 4 months. Directly observed therapy is strongly considered in all patients to limit development of multidrug-resistant organisms in noncompliant patients. Preventative measures in immunocompromised individuals are more aggressive.

4. The acid-fast smear plays an important diagnostic role in mycobacterial infections. The ability to retain aryl methane dyes (carbolfuchsin and auramine O) within the lipid-rich mycobacterial cell wall, after washing with alcohol or weak acids, is a primary feature of mycobacteria. Both Ziehl-Neelsen and Kinyoun methods take advantage of this property to allow quick identification of slowly growing organisms in sputum samples.

CASE STUDY 13

1. Sexual transmission is the primary mode for the spread of HSV-2, and it is likely that the patient had an encounter with an infected individual many years ago. At that time, symptoms were most likely limited to herpetic lesions of the genital area. In general, herpesviruses start with acute infection but progress to lifelong latent infection with periodic reactivation of symptomatic infection.

2. HSV-2 is known for its neurovirulent properties, with latency established in nerve root ganglia. Aseptic meningitis is the main neurologic complication of reactivation of HSV-2 infection. Although herpes-induced encephalitis may occur, it is relatively rare. Rather, herpes-related meningitis occurs in a low but significant number of cases, stemming from primary genital HSV-2. The meningitis is self-limited (lasting less than a week) with only rare persisting neurologic impairment. HSV-2 may also cause benign recurrent lymphocytic meningitis, with lingering presence of elevated activated immune cells and occasional presence of Mollaret cells (giant epithelioid cells of mononuclear origin) in cerebrospinal fluid. Neuralgias are also associated with reactivation infection.

3. Latent HSV can be reactivated by fever, trauma, emotional stress, excessive sunlight, and even menstruation in females. Notably, reactivation is more frequent and severe in immunocompromised patients owing to lack of T_H cell and cytotoxic cell control during early reactivation events. Systemic viral diseases may also trigger general gastroenteritis, possibly due to cytokine cascades released during aggressive T-cell responses.

CASE STUDY 14

1. Because the patient is immunocompromised, with low CD4+ cell numbers, it is likely that an opportunistic infection is the causative agent for the gastrointestinal pathology. Some of the more common agents include *Cryptosporidium*, *Histoplasma*, *Entamoeba histolytica* (amebiasis), and *Campylobacter* spp. In addition, complications due to disseminated *Mycobacteria* cannot be ruled out; a negative purified protein derivative test result may indicate loss of CD4+ cell function rather than lack of exposure.

2. Treatment depends on identification of the opportunistic infection. Because of the high likelihood of a fungal agent, administration of intravenous amphotericin B is warranted, followed by maintenance therapy with oral itraconazole or fluconazole. For *Cryptosporidium*, paromomycin and azithromycin are recommended. For general bacterial infections in the lower intestine, clarithromycin and amoxicillin are first choices unless a drug resistant strain is detected. In all cases, improvement of the immune system and decrease in HIV viral loads (through use of antiviral therapy) will lead to long-lasting results.

3. HAART represents guidelines established to reduce HIV viral burdens to ultimately limit the occurrence of opportunistic infections in immunocompromised individuals. A complex mixture of antiretroviral drugs are used that act at different stages of the HIV life cycle. These drugs include reverse transcriptase inhibitors (both nucleoside and non-nucleoside inhibitors), nucleotide analog reverse transcriptase inhibitors, protease inhibitors, and cell fusion inhibitors, which limit entry of virus into cells.

CASE STUDY 15

1. Prematurity and surgery are the primary predisposing risk factors for candidiasis. Neonates with weakened immune function or defects in mounting a cellular immune response show decreased resistance to fungal infections. Likewise, patients with defects in neutrophil function are also at risk for disseminated disease.

2. Nosocomial acquisition is the most likely form of transmission as a result of contamination of the disposable vascular device (the percutaneously inserted central line). The point of entry for the organism would be the catheter; vascular lines should be monitored to ensure cleanliness to limit infections. Neonatal candidiasis may also occur by vertical transmission (from mother to child during birth), although symptoms would most likely have been manifested earlier in this case.

3. Current therapeutic treatment involves rapid intravenous administration of amphotericin B (or another antifungal agent). If candidemia is not treated immediately, its potential to cause extensive tissue invasion is increased, possibly leading to osteomyelitis, endocarditis, or meningitis. Invasive candidemia may also affect bones and joints and kidney tissue, leading to renal insufficiency. In the majority of cases, catheter-related candidemia will resolve soon after catheter removal and initiation of therapy.

CASE STUDY 16

1. Differential diagnoses include infectious and neoplastic diseases of the liver and gallbladder disease (hence the yellowish sclera). Specific etiologies are hepatocellular cancer or intestinal/hepatic parasitic infection (e.g., *Schistosoma mansoni* or *Strongyloides stercoralis*). Esophagogastroduodenoscopic biopsy should be performed. In this case, histologic examination of collected tissue revealed an intense infiltrate of polymorphonuclear leukocytes, with a predominance of eosinophils. *S. stercoralis* was positively identified.

2. Although *S. stercoralis* is an intestinal nematode, larvae occasionally spread to the respiratory tract, resulting in clinical symptoms ranging from cough to hemoptysis and pulmonary infiltrates. The filariform larvae of *S. stercoralis* may also be encountered in respiratory specimens but are more commonly seen in superinfection in immunosuppressed individuals.

3. *Strongyloides* may be treated with albendazole or thiabendazole. However, it should be noted that anthelmintic therapy works poorly against larvae; treatment must therefore wait for established infection. If *S. stercoralis* is diagnosed in early stages (during larval migration to lung tissue), symptomatic treatment with inhaled β-agonists is sufficient; corticosteroid therapy is not recommended because it leads to compromise of cell-mediated immunity, allowing massive proliferation and widespread dissemination of larvae.

Index

Note: Page numbers followed by *b* indicate boxes, *f* indicate figures, and *t* indicate tables.

A

ABO blood groups, 78–79, 78*f*
Abscess, 109*b*
Acanthamoeba, 151
Acid-fast organisms, 106*t*, 118–120
　intestinal coccidia, 120
　Mycobacterium, 58, 58*f*, 82*t*, 83, 84*f*, 85*b*, 87, 87*f*, 119–120, 119*f*
　Mycoplasma, 120
　Nocardia, 120
Acinetobacter, 115–117
Acquired immune system. *See* Adaptive immune system
Actinomyces, 112
Acute inflammation, 3*b*
Acute respiratory distress syndrome (ARDS), 44*b*
Adaptive immune system, 3–4, 4*f*, 4*t*, 7–8, 53–59, 59*b*
　complement enhancement of, 50
　CTL development in, 54, 56*f*
　hypersensitivity and, 54–59, 59*b*
　　type I: IgE mediated immediate, 54–55, 56*f*
　　type II: antibody-mediated, 55–56, 57*f*
　　type III: immunocomplex-mediated, 56–58, 57*f*
　　type IV: delayed-type, 58–59, 58*f*
　innate immune system links with, 46–47, 48*f*
　specificity of, 4–5, 5*f*
　T cells as primary effectors for cell-mediated immunity, 53, 53*b*
　T cells in B-cell activation, 53–54, 55*f*
　T helper 1 and 2 cell phenotype development in, 53, 53*b*, 54*f*
ADCC. *See* Antibody-dependent cell-mediated cytotoxicity
Adenosine triphosphate-mediated intracellular transporters, 36*b*
Adenovirus, 129
Adjuvants, 69–70, 70*t*
Adrenaline. *See* Epinephrine
Adult bone marrow, 12
Aerobic gram-negative bacilli, 106*t*, 112–114
　Bordetella, 114

Aerobic gram-negative bacilli (*Continued*)
　Campylobacter and *Helicobacter*, 115
　Enterobacteriaceae, 112–113, 114
　Haemophilus, 114
　Legionella, 114
　nonfermenters, 115–117
　zoonotic-origin organisms, 114–115, 116*f*
Aerobic gram-positive bacilli, 106*t*, 109–111
　Bacillus, 109–110, 110*f*
　Corynebacterium, 111
　Erysipelothrix, 111
　Lactobacillus, 110
　Listeria, 110–111
Affinity. *See* Functional affinity
Affinity constants, 71*b*
Affinity maturation, 23, 23*f*, 24*b*
Aflatoxins, 145–146, 146*f*
Allergen, 54–55, 56*f*
Allergen-induced bronchoconstriction, 141*b*
Allergic reactions, 54–55, 56*f*, 141, 141*b*
Allograft, immune-mediated rejection of, 65–66, 67*f*, 68*t*
Alternative pathway of complement activation, 48, 49, 50*f*
Alveolar macrophages, 44–46
Amebae, 148–149, 149*f*
Anaerobic gram-negative bacilli, 106*t*, 117
　Bacteroides, 117
　Fusobacterium, 117
Anaerobic gram-positive bacilli, 106*t*, 111–112
　Actinomyces, 112
　Bifidobacterium, 112
　Clostridium, 111–112, 112*f*
　Eubacterium and *Propionibacterium*, 112
Anaphylactic hypersensitivity, 54–55, 56*f*
Anatomic barrier, 43
Anergy, 62, 63*f*
Animal models of human disease, 78
Anthrax, 109, 110*f*
Antibacterial vaccines, 21*b*
Antibody, 10
　genetics of, 21–22, 22*f*, 24*b*
　　diversity arising from, 22–23, 24*b*

Antibody (*Continued*)
　isotype switching and affinity maturation of, 23, 23*f*, 24*b*
　structure and function of, 17–18, 18*b*, 18*f*, 19*f*, 19*t*, 20*t*
Antibody-antigen interactions, 18–20, 21*b*, 71
　antigen binding, 18, 20*f*
　in immunoassays
　　as basis of quantitative and qualitative assays, 71–72, 72*f*
　　monoclonal and polyclonal antibody specificity for antigen detection, 73–75, 74*f*, 75*f*, 76*f*
　　polyclonal antibody and multivalent antigen interactions, 71
　physiochemical forces in, 18–20, 20*f*
Antibody-antigen precipitation reactions, in gels, 72–73, 73*f*
Antibody-dependent cell-mediated cytotoxicity (ADCC), 47, 49*f*, 55–56
Antibody-mediated hypersensitivity, 55–56, 57*f*
Antifungal agents, 141, 141*f*
Antigen, 3–4. *See also* Antibody-antigen interactions
　immunogens and, 21, 21*b*
　lymphocyte receptors for, 4–5, 5*f*
　nonclassical lipid, 40–41, 41*f*
Antigen presentation
　atypical and unique, 39–41, 41*b*
　　nonclassical lipid antigens, 40–41, 41*f*
　　superantigens, 39–40, 40*f*
　MHC role in, 35, 35*f*, 36*f*
　　endogenous antigen processing, 36–37, 37*f*, 38*b*
　　exogenous antigen processing, 37–38, 38*b*, 38*f*
Antigenic determinant, 18
Antigenic modulation, 87–88
Antigen-presenting cells (APCs), 3–4, 4*f*, 5*f*, 37–38, 61
Antigen-specific receptors, 25–26, 25*b*, 26*f*, 27*f*
Antimicrobial agents, 102–103, 102*t*, 103*b*
Antiviral agents, 124–128, 126*b*
Antiviral vaccination, 125–128, 127*t*

Index

APCs. *See* Antigen-presenting cells
Apoptosis, 28*b*, 62, 63*b*, 63*f*
   CTL induction of, 29–31
   in T cell development, 27–28
Appendix, 15, 15*f*
Arboviruses, 130*t*, 135–136
   dengue virus, 135
   Japanese encephalitis virus, 135
   tick-borne encephalitis, 135
   VSV, 135
   West Nile virus, 135
   yellow fever, 135
ARDS. *See* Acute respiratory distress syndrome
Arthus reaction, 56–58
Ascaris, 85, 151
Aspergillus, 142*f*, 145–146, 146*f*
Assays. *See* Immunoassays
Atypical mycobacteria, 119
Autograft, 65
Autoimmune vasculitis, 58*b*
Autoimmunity
   etiology of, 63–65, 64*t*, 65*b*, 66*f*
   MHC haplotypes associated with, 39, 39*b*
Avidity. *See* Functional affinity
Azathioprine, 64*t*

B

B cell, 3–4, 4*f*, 10, 10*t*
   antigen presentation by, 38
   antigen receptors of, 4–5, 5*f*
   deficiency of, 68*f*
   function assays, 76–77, 77*f*
   in humoral immunity, 7
   lymphoid lineage of, 7–8, 7*b*, 10, 10*t*
   organs producing, 12
   self-reactive, 62–63
   T cell role in activation of, 53–54, 55*f*
   tolerance of, 62, 63*f*
Babesia, 150
Bacilli. *See* Gram-negative bacilli; Gram-positive bacilli
Bacillus, 109–110, 110*f*
Bacteremia, 100
Bacterial infection, 91–103, 103*b*, 105–120, 106*t*, 120*b*
   acid-fast organisms, 106*t*, 118–120
      intestinal coccidia, 120
      Mycobacterium, 58, 58*f*, 82*t*, 83, 84*f*, 85*b*, 87, 87*f*, 119–120, 119*f*
      Mycoplasma, 120
      Nocardia, 120
   aerobic gram-negative bacilli, 106*t*, 112–114
      Bordetella, 114
      Campylobacter and *Helicobacter*, 115
      Enterobacteriaceae, 112–113, 114
      Haemophilus, 114
      Legionella, 114
      nonfermenters, 115–117
      zoonotic-origin organisms, 114–115, 116*f*

Bacterial infection (*Continued*)
   aerobic gram-positive bacilli, 106*t*, 109–111
      Bacillus, 109–110, 110*f*
      Corynebacterium, 111
      Erysipelothrix, 111
      Lactobacillus, 110
      Listeria, 110–111
   anaerobic gram-negative bacilli, 106*t*, 117
      Bacteroides, 117
      Fusobacterium, 117
   anaerobic gram-positive bacilli, 106*t*, 111–112
      Actinomyces, 112
      Bifidobacterium, 112
      Clostridium, 111–112, 112*f*
      Eubacterium and *Propionibacterium*, 112
   antimicrobial agents for, 102–103, 102*t*, 103*b*
   bacteremia, SIRS, and sepsis, 100
   clinical diagnosis of, 102
   commensal organisms of normal body flora, 95–98, 97*f*, 98*b*
   genetics of bacteria, 98–100, 100*b*
      gene expression and regulation, 100
      genetic transfer methods, 98–100, 98*f*, 99*f*
   gram-negative cocci, 106*t*, 108–109
      Neisseria, 108–109
      Veillonella, 109
   gram-positive cocci, 105–108, 106*t*, 107*f*
      staphylococci, 105–107, 107*f*
      streptococci and enterococci, 107–108, 110*f*
   major immune defense mechanisms against, 81–82, 82*t*, 83*f*
   obligate intracellular pathogens, 106*t*, 117–118
      Chlamydia, 117–118, 118*f*
      Rickettsia, *Ehrlichia*, and *Coxiella*, 118
   physiology of bacteria, 95, 95*b*, 96*f*
   prokaryotic properties, 93–95, 94*f*
   spirochetes, 106*t*, 117
      Borrelia, 117
      Leptospira, 117
      Treponema, 75–76, 117
   structure, function, and classification of bacteria, 93, 94*f*, 95*b*
   toxins as virulence factors, 100–102, 100*f*, 101*t*, 102*b*
Bacteriophages, 98
Bacteroides, 117
Balantidium, 149
BALT. *See* Bronchus-associated lymphoid tissue
Baltimore Classification of viruses, 123, 125*t*
Barriers, against infection, 43–44, 44*b*
Bartonella, 115
Basophils, 8*t*, 9
B-cell receptor, 29*f*

Bifidobacterium, 112
BK viruses, 135
Blast transformation assay, 76–77, 77*f*
Blastomyces, 145
Body flora, normal, 95–98, 97*f*, 98*b*
Bone marrow, 12
Bordetella, 114
Borrelia, 117
Botulism, 111, 112*f*
Bradykinin, 9*b*
Bronchoconstriction, allergen-induced, 141*b*
Bronchus-associated lymphoid tissue (BALT), 15
Brucella, 115
Bubonic plague, 113
Burkitt lymphoma, 134, 134*b*

C

C3 convertase, 49
Campylobacter, 115
Cancer
   immune therapy for, 69–70, 70*t*
   immunology of, 66, 66*b*
Candida, 85–86, 142, 143, 145
CD. *See* Cluster of differentiation
CD1, 40–41, 41*f*
CD3, 10, 25–26, 26*f*
CD4, 10, 25–26, 26*f*, 27–29, 30*f*, 31–32
   in MHC antigen presentation, 35, 37–38
CD8, 10–11, 25–26, 26*f*, 27–31, 30*f*
   in MHC antigen presentation, 35, 35*f*, 36–37
CD25, 31–32
CDRs. *See* Complementarity-determining regions
Cell-mediated hypersensitivity, 58–59, 58*f*
Cellular immunity, 7, 53, 53*b*
Central lymphoid organs. *See* Primary lymphoid organs
Cestodes, 152–153, 153*b*, 153*f*
Chagas disease, 150–151
Chemokines, 44–46, 46*t*
Chickenpox, 131–132
Childhood exanthems, 130*t*, 131–132
   mumps, 131
   parvovirus, 131
   rubella, 131
   rubeola, 131
   varicella, 131–132
Chlamydia, 117–118, 118*f*
Cholera, 115, 116*f*
Chromomycosis, 144
Chromosomes, 99*b*
Cilia, respiratory, 96*b*
Ciliates, 149
Class I MHC molecules, 33–34, 34*b*, 34*f*
   structure of, 34–38, 35*b*, 35*f*
      antigen presentation and, 35
      endogenous antigen processing and, 36–37, 37*f*, 38*b*

Class II MHC molecules, 33–34, 34b, 34f
　structure of, 34–38, 35b, 36f
　　antigen presentation and, 35
　　exogenous antigen processing and, 37–38, 38b, 38f
Classic pathway of complement activation, 48–49, 50f
Clathrin-mediated endocytosis, 123b
Clinical bacteriology. *See* Bacterial infection
Clinical mycology. *See* Fungal infection
Clinical virology. *See* Viral infection
Clonal selection, 4–5
Clostridium, 111–112, 112f
Cluster of differentiation (CD), 10b
Cocci. *See* Gram-negative cocci; Gram-positive cocci
Coccidia, 149
　intestinal, 120
Coccidioides, 145
Codominant gene expression, 33–34, 34b
Colon, 149b
Combinatorial diversity, 22
Commensal organisms, of normal body flora, 95–98, 97f, 98b
Complement, 47–50, 50b
　disorders of, 68f
　in inflammation and immunity
　　biologic functions, 49–50, 51f
　　cascades, 48–49, 50f
Complement fixation test, 75–76, 77f
Complementarity-determining regions (CDRs), 22, 25–26, 27f
Conformational epitope, 21
Conjugation, 98, 98f
Coombs test, 72, 72f
Coronavirus, 131
Corticosteroids, 64t, 82b
Corynebacterium, 111
Coxiella, 118
Coxsackievirus, 132
C-reactive protein (CRP), 75
Creutzfeldt-Jakob disease, 136
Cross-reactivity, 21
CRP. *See* C-reactive protein
Cryptococcus, 145
Cryptosporidium, 149
CTLs. *See* Cytotoxic T lymphocytes
Cutaneous larva migrans, 152b
Cutaneous mycoses, 143–144, 143f
Cyclospora, 149
Cyclosporine, 64t
Cytolysis, of foreign organisms, 49
Cytomegalovirus. *See* Human cytomegalovirus
Cytoplasmic antigen processing, 36–37, 37f, 38b
Cytotoxic antibody, 55
Cytotoxic T lymphocytes (CTLs), 10–11
　development of, 54, 56f
　functions of, 29–31, 31b, 31f

D
D genes. *See* Diversity genes
DCs. *See* Dendritic cells
Delayed-type hypersensitivity (DTH), 58–59, 58f
Dendritic cells (DCs), 46–47, 48f
　myeloid lineage of, 8t, 9
Dengue virus, 135
Diarrhea, 110b, 112
Dientamoeba, 149
Diphtheria, 111
Diversity (D) genes, recombination of
　for antibodies, 21–23, 22f, 24b
　for TCRs, 26–27, 26b, 28f, 29f
DNA, viral, 121–122, 123f
Double diffusion assay, 72–73, 73f
DTH. *See* Delayed-type hypersensitivity
Dysentery, 148–149, 149f

E
EBV. *See* Epstein-Barr virus
Echinococcus, 152
Echoviruses, 132
Ehrlichia, 118
ELISA. *See* Enzyme-linked immunosorbent assay
Encephalitis. *See* Japanese encephalitis virus; Tick-borne encephalitis
Endocytic barrier, 43–44
Endocytosis, 123b
Endogenous antigen processing, 36–37, 37f, 38b
Endosomal antigen processing, 37–38, 38b, 38f
Endotoxins, 100–102, 100f, 101t, 102b
Enterobacteriaceae, 112–113, 114
　Escherichia coli, 113
　Salmonella, 113
　Shigella, 113
　Yersinia, 113
Enterococci, 107–108, 110f
Enterotoxins, 100–102, 101t
Enteroviruses, 130t, 132
　coxsackievirus, 132
　echoviruses, 132
　poliovirus, 132
Enzyme-linked immunosorbent assay (ELISA), 73–74, 75f
Eosinophils, 8t, 9
Epidermophyton, 143
Epinephrine (Adrenaline), 56b
Epitope, 3–4, 18, 71
　conformational, 21
　cross-reactivity and, 21
　sequential, 21
Epstein-Barr virus (EBV), 88–89, 133t, 134
Erysipelothrix, 111
Erythema migrans, 117b
Erythroblastosis fetalis, 78–79
Erythrocyte. *See* Red blood cells
Erythrocyte sedimentation rate (ESR), 75
Escherichia coli, 113
ESR. *See* Erythrocyte sedimentation rate
Eubacterium, 112
Eukaryotes, 148b
Exogenous antigen processing, 37–38, 38b, 38f
Exophiala werneckii, 143
Exotoxins, 100–102, 100f, 101t, 102b
Extraintestinal protozoa, 149–151, 150f

F
FACS. *See* Fluorescence-activated cell sorting
Fetal liver, 12
Fever, 50–51
Fifth disease, 131
Filariae, 152
Flagellates, 149, 149f
Flu. *See* Influenza
Fluorescence-activated cell sorting (FACS), 74–75, 76f
Food poisoning, 109–110
Foreign organisms. *See* Pathogens
Foxp3 mutations, 31–32
Francisella, 115
Functional affinity, 18, 20f, 71–72
Fungal infection, 139–146, 146b
　antifungal agents, 141, 141f
　clinical, 143–146
　　subcutaneous mycoses, 144
　　superficial and cutaneous mycoses, 143–144, 143f
　　systemic mycoses, 144–146, 144f, 146f
　fungal pathogenesis, 139–141
　fungal structure and classification, 139, 140f, 140t
　immune response to fungal agents, 142, 142f
　laboratory diagnosis of fungal infections, 141–142
　major immune defense mechanisms against, 82t, 85–86
Fusobacterium, 117

G
α-Galactosylceramide, 32, 69–70
GALT. *See* Gut-associated lymphoid tissue
Gas gangrene, 111–112
Gels, antibody-antigen precipitation reactions in, 72–73, 73f
Gene chips, 77–78
Gene transcription, 22b
Genetics
　of antibody structure, 21–22, 22f, 24b
　　diversity arising from, 22–23, 24b
　bacterial, 98–100, 100b
　　gene expression and regulation, 100
　　genetic transfer methods, 98–100, 98f, 99f
　of MHC, 33–34, 34b, 34f
　of TCRs, 26–27, 26b, 28f, 29f
　of viruses, 121–122, 123f

Genital herpes, 132–133, 133t, 134f
German measles, 131
Gerstmann-Straussler disease, 136
Giardia, 149, 149f
Gonococcal infection, 108–109
Graft tissue, immune-mediated rejection of, 65
Graft-*versus*-host disease (GVHD), 66, 66b, 67f
Gram-negative bacilli
　aerobic, 106t, 112–114
　　Bordetella, 114
　　Campylobacter and *Helicobacter*, 115
　　Enterobacteriaceae, 112–113, 114
　　Haemophilus, 114
　　Legionella, 114
　　nonfermenters, 115–117
　　zoonotic-origin organisms, 114–115, 116f
　anaerobic, 106t, 117
　　Bacteroides, 117
　　Fusobacterium, 117
Gram-negative cocci, 106t, 108–109
　Neisseria, 108–109
　Veillonella, 109
Gram-negative organisms, 93, 94f
Gram-positive bacilli
　aerobic, 106t, 109–111
　　Bacillus, 109–110, 110f
　　Corynebacterium, 111
　　Erysipelothrix, 111
　　Lactobacillus, 110
　　Listeria, 110–111
　anaerobic, 106t, 111–112
　　Actinomyces, 112
　　Bifidobacterium, 112
　　Clostridium, 111–112, 112f
　　Eubacterium and *Propionibacterium*, 112
Gram-positive cocci, 105–108, 106t, 107f
　staphylococci, 105–107, 107f
　streptococci and enterococci, 107–108, 110f
Gram-positive organisms, 93, 94f
Granuloma, 83, 85b, 87
Group A streptococci, 107
Group B streptococci, 21b, 107–108
Gut-associated lymphoid tissue (GALT), 15
GVHD. *See* Graft-*versus*-host disease

H
HAART. *See* Highly active antiretroviral therapy
Haemophilus, 114
Hansen disease. *See* Leprosy
Haptens, 21
HCMV. *See* Human cytomegalovirus
Heavy chains
　of antibodies, 17–18, 18f, 19t
　recombination of, 21–23, 24b
Helicobacter, 115

Helminths, 82t, 85, 86f. *See also* Parasitic infection
Helper T cell. *See* T helper cell
Hepatitis viruses, 130t, 132, 133f
Hepatosplenomegaly, 145b
Herpes virus, 88–89, 130t, 132–134, 133t
　cytomegalovirus, 133–134, 133t
　EBV, 88–89, 133t, 134
　HHSV1 and HHSV2, 130t, 132–133, 133t, 134f
　HHSV7 and HHSV8, 133t, 134
　HHV6, 131–132, 133t
Herpes zoster, 131–132
Heterograft, 65
HHSV1. *See* Human herpes simplex virus type 1
HHSV2. *See* Human herpes simplex virus type 2
HHSV7. *See* Human herpes simplex virus type 7
HHSV8. *See* Human herpes simplex virus type 8
HHV6. *See* Human herpes simplex virus type 6
Highly active antiretroviral therapy (HAART), 126b
Hinge region, 17, 18b
Histocytes, 44–46
Histones, 99b
Histoplasma, 142f, 145, 145b
HIV. *See* Human immunodeficiency virus
Hives. *See* Urticaria
HLA locus. *See* Human leukocyte antigen locus
Homograft. *See* Allograft
Hookworm, 151–152
HPVs. *See* Human papillomaviruses
HTLV. *See* Human T lymphocyte virus
Human cytomegalovirus (HCMV), 133–134, 133t
Human herpes simplex virus type 1 (HHSV1), 132–133, 133t, 134f
Human herpes simplex virus type 2 (HHSV2), 132–133, 133t, 134f
Human herpes simplex virus type 6 (HHV6), 131–132, 133t
Human herpes simplex virus type 7 (HHSV7), 133t, 134
Human herpes simplex virus type 8 (HHSV8), 133t, 134
Human immunodeficiency virus (HIV), 73, 74f, 135, 136f, 137f
Human leukocyte antigen (HLA) locus, 33–34, 34b, 34f
Human papillomaviruses (HPVs), 134–135, 134t
Human T lymphocyte virus (HTLV), 135
Humoral immunity, 7, 17–24
　antibodies, 10
　genetics of, 21–23, 22f, 24b
　isotype switching and affinity maturation, 23, 23f, 24b

Humoral immunity (*Continued*)
　　structure and function of, 17–18, 18b, 18f, 19f, 19t, 20t
　antibody-antigen interactions, 18–20, 21b
　　antigen binding, 18, 20f
　　physiochemical forces in, 18–20, 20f
　antigens and immunogens, 21, 21b
Hydatid worm, 152
Hypersensitivity, 54–59, 59b, 83
　type I: IgE mediated immediate, 54–55, 56f
　type II: antibody-mediated, 55–56, 57f
　type III: immunocomplex-mediated, 56–58, 57f
　type IV: delayed-type, 58–59, 58f

I
Idiotype, 18
IFNs. *See* Interferons
Ig. *See* Immunoglobulin
Immediate hypersensitivity, 54–55, 56f
Immune defense mechanisms, against infection, 81–86, 82f, 82t, 83f, 86b
　bacterial infections, 81–82, 82t, 83f
　fungal infections, 82t, 85–86
　mycobacterial infections, 82t, 83, 84f, 85b
　parasitic infections, 82t, 85, 86f
　viral infections, 82t, 83–85, 84f
Immune deviation, 86–88, 87b, 87f
Immune dysfunction, 5, 6f
Immune response
　evasion of, 87–89, 88b, 88f, 88t
　to fungal agents, 142, 142f
　to parasitic infections, 154–155, 154b
Immune system, 1–6, 6b
　adaptive (*See* Adaptive immune system)
　cells of, 7–16, 16b
　　lymphoid cells, 7–8, 7b, 10–11, 10t, 11b
　　myeloid cells, 7–8, 7b, 8–9, 8t, 9b
　　pluripotent hematopoietic stem cells, 7–8, 8f, 8t
　functional assessment of, 75–79
　　ABO blood groups and Rh incompatibility, 78–79, 78f
　　animal models, 78
　　complement fixation test, 75–76, 77f
　　general inflammation assessment, 75
　　lymphocyte function assays, 76–77, 77f
　　microarrays to assess gene expression, 77–78
　innate (*See* Innate immune system)
　organs of, 7–16, 16b
　　primary lymphoid organs, 11–16, 11f, 12f, 15b
　　secondary lymphoid organs, 11–16, 11f, 13f, 14f, 15b, 15f
　regulation of, 5, 6f
Immune therapy, for cancer, 69–70, 70t
Immunity
　adaptive (*See* Adaptive immune system)
　cellular, 7, 53, 53b

Immunity (*Continued*)
  chief function of, 3
  complement in, 47–50, 50*b*
    biologic functions of, 49–50, 51*f*
    cascades, 48–49, 50*f*
  humoral (*See* Humoral immunity)
  innate (*See* Innate immune system)
  T cell, 25–32, 32*b*
    antigen-specific receptors in, 25–26, 25*b*, 26*f*, 27*f*
    T cell development in, 27–29, 30*f*
    T cell phenotype functions in, 29–32, 29*b*, 30*f*, 31*b*, 31*f*, 31*t*, 32*b*
    TCR genes in, 26–27, 26*b*, 28*f*, 29*f*
Immunization. *See* Vaccination
Immunoassays, 71–79, 79*b*
  antibody-antigen interactions in
    as basis of quantitative and qualitative assays, 71–72, 72*f*
    monoclonal and polyclonal antibody specificity for antigen detection, 73–75, 74*f*, 75*f*, 76*f*
    polyclonal antibody and multivalent antigen interactions, 71
  antibody-antigen precipitation reactions in gels, 72–73, 73*f*
  for immune function assessment, 75–79
    ABO blood groups and Rh incompatibility, 78–79, 78*f*
    animal models, 78
    complement fixation test, 75–76, 77*f*
    general inflammation assessment, 75
    lymphocyte function assays, 76–77, 77*f*
    microarrays to assess gene expression, 77–78
Immunoblotting, 73, 74*f*
Immunocomplexes, solubilization and clearance of, 50
Immunocomplex-mediated hypersensitivity, 56–58, 57*f*
Immunodeficiency, 5, 6*f*, 67, 67*b*, 68*f*. *See also* Human immunodeficiency virus
Immunogens, 21, 21*b*
Immunoglobulin (Ig), 17–18
  IgA, 17, 18, 19*f*, 19*t*
  IgD, 17, 18, 19*f*, 19*t*
  IgE, 17, 18, 19*f*, 19*t*
  IgG, 17, 18, 19*f*, 19*t*, 20*f*, 20*t*
  IgM, 17, 18, 19*f*, 19*t*, 20*f*
Immunoglobulin E-mediated immediate hypersensitivity, 54–55, 56*f*
Immunohistochemical assays, 74, 76*f*
Immunomodulation, 61–70, 70*b*
  autoimmune disease and, 63–65, 64*t*, 65*b*, 66*f*
  in cancer, 66, 66*b*, 69–70, 70*t*
  immunodeficiency as predisposition to infection, 67, 67*b*, 68*f*
  immunoprophylaxis against infection, 68–69, 69*b*, 69*f*, 69*t*
  T cell activation events, 61, 61*b*, 62*f*
  tolerance in, 61–63, 63*b*

Immunomodulation (*Continued*)
  lymphocyte tolerance, 62, 63*f*
  self-reactive lymphocyte elimination, 62–63
  transplantation rejection and, 65–66, 66*b*, 67*f*, 68*t*
Immunoprophylaxis, 68–69, 69*b*, 69*f*, 69*t*
Immunosuppression, 87–88
Immunosuppressive agents, 65–66, 68*t*
Infection, 81–89, 89*b*. *See also specific types*
  evasion of immune response and, 87–89, 88*b*, 88*f*, 88*t*
  immune deviation and, 86–88, 87*b*, 87*f*
  immunodeficiency as predisposition to, 67, 67*b*, 68*f*
  immunoprophylaxis against, 68–69, 69*b*, 69*f*, 69*t*
  major immune defense mechanisms against pathogens, 81–86, 82*f*, 82*t*, 83*f*, 86*b*
    bacterial infections, 81–82, 82*t*, 83*f*
    fungal infections, 82*t*, 85–86
    mycobacterial infections, 82*t*, 83, 84*f*, 85*b*
    parasitic infections, 82*t*, 85, 86*f*
    viral infections, 82*t*, 83–85, 84*f*
  MHC haplotypes associated with, 39, 39*b*
  physical barriers against, 43–44, 44*b*
  viral strategies for, 122–123, 123*b*, 124*f*, 125*t*
Inflammation
  complement in, 47–50, 50*b*
    biologic functions of, 49–50, 51*f*
    cascades, 48–49, 50*f*
  general assessment of, 75
  mediators of, 3*b*
  pharmacologic control of, 82*b*
Inflammatory barriers, 44, 44*f*
Influenza, 114, 130
Innate immune system, 3, 4*t*, 7–8, 43–51, 51*b*
  complement, 47–50, 50*b*, 68*f*
    biologic functions of, 49–50, 51*f*
    cascades, 48–49, 50*f*
  defensive responding cells of, 44–46, 46*b*
    macrophages, 8*t*, 9, 44–46, 46*b*, 46*t*, 47*f*
    mononuclear cells, 44–46, 46*t*
    neutrophils, 8–9, 8*t*, 44, 45*f*
  fever and rash in, 50–51
  links between adaptive immunity and, 46–47, 48*f*
  microbial agent recognition through PRRs, 46, 46*b*, 47*f*
  NK cells, 10*t*, 11, 47, 49*f*
  physical barriers against infection, 43–44, 44*b*
Insertional diversity, 23
Interferons (IFNs), 125
Intestinal infection
  Coccidia, 120
  Enterobacteriaceae, 112–113
  protozoan, 148–149, 149*f*

Intestine, 113*b*
Isograft, 65
Isohemagglutinins, 78
Isospora, 149
Isotype, 17, 19*f*, 19*t*
Isotype switching, 23, 23*f*, 24*b*

J

J genes. *See* Joining genes
Japanese encephalitis virus, 135
JC viruses, 135
Joining (J) genes, recombination of
  for antibodies, 21–23, 22*f*, 24*b*
  for TCRs, 26–27, 26*b*, 28*f*, 29*f*
Jumping genes. *See* Transposons
Junctional diversity, 23

K

Kaposi sarcoma, 134
Keratinocytes, 134–135, 134*b*
Kupffer cells, 44–46
Kuru, 136

L

β-Lactams, 102–103, 103*b*
Lactobacillus, 110
Lectin pathway of complement activation, 48, 50*f*
Legionella, 114
Leishmania, 151
Leprosy, 82*t*, 83, 84*f*, 87, 87*f*, 119–120
Leptospira, 117
Leukocytes, 7–8, 8*f*, 49. *See also* Lymphoid cells; Myeloid cells
Leukotrienes, 3*b*, 9*b*
Lice, 153, 153*f*
Light chains
  of antibodies, 17–18, 18*f*
  recombination of, 21–23, 24*b*
Lipopolysaccharide (LPS), 94, 100–102
Listeria, 110–111
Liver, 12, 153*b*
LPS. *See* Lipopolysaccharide
LSD. *See* Lysergic acid diethylamide
Lyme disease, 117
Lymph, 14
Lymph nodes, 14, 14*f*
Lymphatic vessels, 14, 144*b*
Lymphocyte. *See also* B cell; T cell
  antigen receptors of, 4–5, 5*f*
  function assays, 76–77, 77*f*
  self-reactive, 62–63
  tolerance of, 62, 63*f*
Lymphoid cells, 7–8, 7*b*, 10–11, 10*t*, 11*b*
  B cells, 7–8, 7*b*, 10, 10*t*
  NK cells, 10*t*, 11
  NKT cells, 10*t*, 11
  ontogenic development of, 11*b*
  T cells, 7–8, 7*b*, 10–11, 10*t*

Lymphoid organs
  primary, 11–16, 11f, 15b
    fetal liver and adult bone marrow, 12
    thymus gland, 12–13, 12f, 39b
  secondary, 11–16, 11f, 15b
    lymph nodes, 14, 14f
    MALT, 15–16, 15f
    spleen, 13–14, 13f
Lymphoreticular system.
    See Reticuloendothelial system
Lysergic acid diethylamide (LSD), 139–141, 141b
Lyssavirus, 134

M
Macrophages, 44–46, 46b, 46t, 47f
  myeloid lineage of, 8t, 9
Major histocompatibility complex (MHC), 3–4, 4f, 33–41, 41b
  in atypical and unique antigen presentation, 39–41, 41b
    nonclassical lipid antigens, 40–41, 41f
    superantigens, 39–40, 40f
  class I and II molecule structure, 34–38, 35b, 35f, 36f
    antigen presentation and, 35, 35f, 36f
    endogenous antigen processing and, 36–37, 37f, 38b
    exogenous antigen processing and, 37–38, 38b, 38f
  diseases associated with haplotypes of, 39, 39b
  function of, 33
  organization and polymorphism of MHC genes, 33–34, 34b, 34f
  in T cell activation, 39, 39f
  T cell recognition of, 25–26, 27f
  in thymic education, 38, 38b
Malaria, 150, 150f
Malassezia furfur, 143
MALT. See Mucosa-associated lymphoid tissue
Mannose-binding pathway. See Lectin pathway of complement activation
Mantoux tuberculin skin test, 58, 58f
Masking, 87–88
Mast cells, 8t, 9
Measles, 131
Membrane-bound antigens, Western blot for detection of, 73, 74f
Memory, immunologic, 5, 5f
Meningitis, 109
Mesangial cells, 44–46
MHC. See Major histocompatibility complex
Microarrays, 77–78
Microbial agent. See Pathogens
Micrococcus spp, 108
Microglial cells, 44–46
Microsporum, 143
Mites, 153, 153f

Monoclonal antibodies, 10, 69–70, 70t
  specificity for antigen detection in immunoassays, 73–75, 74f, 75f, 76f
Monocytes, 44–46, 46t, 47f
  myeloid lineage of, 8t, 9
Mononuclear cells, 44–46, 46t
Mononuclear phagocytic system.
    See Reticuloendothelial system
Mononucleosis, 134
Mucor, 146
Mucosa-associated lymphoid tissue (MALT), 15–16, 15f
Mucous membranes, 43
Multiple sclerosis, 64b
Multivalent antigens, polyclonal antibody interactions with, 71
Mumps, 131
Mycetoma, 144
Mycobacterial infections, 119–120
  atypical mycobacteria, 119
  leprosy, 82t, 83, 84f, 87, 87f, 119–120
  major immune defense mechanisms against, 82t, 83, 84f, 85b
  tuberculosis, 58, 58f, 82t, 83, 84f, 85b, 119, 119f
Mycology. See Fungal infection
Mycoplasma, 120
Myeloid cells, 7–8, 7b, 8–9, 9b
  basophils, 8t, 9
  DCs, 8t, 9
  eosinophils, 8t, 9
  macrophages, 8t, 9
  mast cells, 8t, 9
  monocytes, 8t, 9
  neutrophils, 8–9, 8t
  platelets and erythrocytes, 9

N
Naegleria, 151
Natural immune system. See Innate immune system
Natural killer (NK) cell, 47, 49f
  lymphoid lineage of, 10t, 11
Natural killer T (NKT) cell
  functions of, 32, 32b
  lymphoid lineage of, 10t, 11
Neisseria, 108–109
Nematodes, 151–152, 151f, 153b
Neutrophils, 44, 45f
  myeloid lineage of, 8–9, 8t
NK cell. See Natural killer cell
NKT cell. See Natural killer T cell
Nocardia, 120
Nonclassical lipid antigens, 40–41, 41f
Noncovalent bond forces, 20f, 21b, 71–72
Nonfermenters, 115–117
  Acinetobacter, 115–117
  Pseudomonas, 115–117
Nonsteroidal antiinflammatory drugs (NSAIDs), 82b
Normal body flora, 95–98, 97f, 98b

NSAIDs. See Nonsteroidal antiinflammatory drugs

O
Obligate intracellular pathogens, 106t, 117–118
  Chlamydia, 117–118, 118f
  Rickettsia, *Ehrlichia*, and *Coxiella*, 118
Omenn syndrome, 26, 27b
Opsonization, of foreign organisms, 49
Osteoclasts, 44–46
Ouchterlony assay, 72–73, 73f

P
Papillomaviruses. See Human papillomaviruses
Papovaviruses, 130t, 134–135
  HPVs, 134–135, 134t
  JC and BK viruses, 135
Paracoccidioides, 145
Parainfluenza virus, 130
Parasitic infection, 147–155, 154b
  cestodes and trematodes, 152–153, 153b, 153f
  immune response to, 154–155, 154b
  laboratory diagnosis of, 153–154, 154b
  lice and mites, 153, 153f
  major immune defense mechanisms against, 82t, 85, 86f
  nematodes, 151–152, 151f, 153b
  pathogenesis of parasites, 147, 147b, 148b
  protozoa, 147–151, 148f, 151b
    extraintestinal, 149–151, 150f
    intestinal, 148–149, 149f
Paratope, 18
Parvovirus, 131
Pasteurella, 114–115
Pathogens. See also specific agents
  direct cytolysis of, 49
  major immune defense mechanisms against, 81–86, 82f, 82t, 86b
    bacterial infections, 81–82, 82t, 83f
    fungal infections, 82t, 85–86
    mycobacterial infections, 82t, 83, 84f, 85b
    parasitic infections, 82t, 85, 86f
    viral infections, 82t, 83–85, 84f
  opsonization of, 49
  PRRs in recognition of, 46, 46b, 47f
Pattern recognition receptors (PRRs), 46, 46b, 47f
Penicillins, 102–103, 103b
Pentose-phosphate pathway, 96b
Peptostreptococci, 108
Pertussis, 114
Peyer patches, 15, 15f
Phagocyte deficiency, 68f
Phagocytic barrier, 43–44
Phagocytosis, 46, 47b, 47f
Phagosome, 46

Physical barriers, 43–44, 44b
Physiochemical forces, in antibody-antigen interactions, 18–20, 20f
Physiologic barrier, 43
Piedraia hortae, 143
Pinocytosis, 122
Pinworms, 151
Plague, 113
Plasma cells, 10
Plasmids, 98
Plasmodium, 150, 150f
Platelets, 9
Pluripotent hematopoietic stem cells, 7–8, 8f, 8t
Pneumocystis, 146
Pneumonia
   fungal, 144–146, 144f
   Legionella, 114
   Streptococcus, 108
Poliovirus, 132
Polyclonal antibodies
   multivalent antigen interactions with, 71
   specificity for antigen detection in immunoassays, 73–75, 74f, 75f, 76f
Polyclonal antilymphocyte serum, 64t
Positive selection, 38
Poxviruses, 130t, 132
Presentation. See Antigen presentation
Primary lymphoid organs, 11–16, 11f, 15b
   fetal liver and adult bone marrow, 12
   thymus gland, 12–13, 12f, 39b
Prions, 130t, 136
Prokaryotes, 93–95, 94f
Prophylaxis. See Immunoprophylaxis
Propionibacterium, 112
Prostaglandins, 3b, 9b
Protective niche, 87–88
Protozoa, 147–151, 148f, 151b
   extraintestinal, 149–151, 150f
   intestinal, 148–149, 149f
PRRs. See Pattern recognition receptors
Pseudomonas, 115–117
Pus, 8–9, 44, 81–82, 109b

R
Rabies, 134
Radial immunodiffusion assay, 73, 73f
RAG-1, 21, 26, 78
RAG-2, 21, 26, 78
Rash, 50–51
RBCs. See Red blood cells
Recombination, genetic
   for antibodies, 21–23, 22f, 24b
   for TCRs, 26–27, 26b, 28f, 29f
Red blood cells (RBCs), 9
Regulation
   bacterial, 100
   of immune system, 5, 6f
Regulatory T cells. See T regulatory cells
Rejection, of transplant, 65–66, 66b, 67f, 68t
Replication

Replication (Continued)
   bacterial, 98–100, 99f
   viral, 122–123, 123b, 124f, 125t
RES. See Reticuloendothelial system
Respiratory burst, 46
Respiratory cilia, 96b
Respiratory syncytial virus (RSV), 131
Respiratory viruses, 129–131, 130t
   adenovirus, 129
   coronavirus, 131
   influenza, 130
   parainfluenza, 130
   respiratory syncytial virus, 131
   rhinovirus, 131
Reticuloendothelial system (RES), 9
Retroviruses, 130t, 135
   HIV, 73, 74f, 135, 136f, 137f
   HTLV, 135
Rh incompatibility, 78–79
Rhabdoviruses, 130t, 134
   Lyssavirus, 134
Rhesus antigens, 78–79
Rhinovirus, 131
RhoGAM, 78–79
Rickettsia, 118
Rift valley fever, 136
RNA, viral, 121–122, 123f
RNA translation, 22b
Roseola infantum, 131–132
Ross River virus, 136
Roundworms. See Nematodes
RSV. See Respiratory syncytial virus
Rubella, 131
Rubeola, 131

S
Salmonella, 113
SARS. See Severe acute respiratory syndrome
Schistosomes, 152–153, 153f
SCID. See Severe combined immunodeficiency
Secondary lymphoid organs, 11–16, 11f, 15b
   lymph nodes, 14, 14f
   MALT, 15–16, 15f
   spleen, 13–14, 13f
Self-reactive lymphocytes, 62–63
Sepsis, 100
Sequential epitope, 21
Severe acute respiratory syndrome (SARS), 131
Severe combined immunodeficiency (SCID), 26, 27b, 78
Shigella, 113
Shingles, 131–132
Shock, 39–40, 40f
Single-positive cells, 38
Sirolimus, 64t
SIRS. See Systemic inflammatory response syndrome
Skin, 43, 43b

Slapped-cheek disease, 131
Smallpox, 132
Solid-phase immunoassays. See Enzyme-linked immunosorbent assay
Spirochetes, 106t, 117
   Borrelia, 117
   Leptospira, 117
   Treponema, 75–76, 117
Spleen, 13–14, 13f
Split tolerance. See immune deviation
Sporotrichosis, 144
Staphylococci, 105–107, 107f
Staphylococcus aureus, 105–107
Stem cells, pluripotent hematopoietic, 7–8, 8f, 8t
Streptococci, 107–108, 110f
Streptomyces, 143–144
Strongyloides, 152
Subcutaneous mycoses, 144
Superantigens, 39–40, 40f
Superficial mycoses, 143–144, 143f
Suppressor T cells. See T suppressor cells
Syngeneic graft, 65
Syphilis, 75–76, 117
Systemic inflammatory response syndrome (SIRS), 100
Systemic mycoses, 144–146, 144f, 146f

T
T cell, 3–4, 4f. See also Natural killer T cell
   activation of
      DCs in, 46–47, 48f
      events involved in, 61, 61b, 62f
      MHC in, 39, 39f
   antigen receptors of, 4–5, 5f
   in B-cell activation, 53–54, 55f
   in cellular immunity, 7
   deficiency of, 68f
   development of, 27–29, 30f
   education of, 38, 38b
   function assays, 76–77, 77f
   lymphoid lineage of, 7–8, 7b, 10–11, 10t
   MHC antigen presentation to, 35, 35f, 36–38, 36f, 37f, 38b, 38f
   organs producing, 12–13
   as primary effectors for cell-mediated immunity, 53, 53b
   self-reactive, 62–63
   tolerance of, 62, 63f
$\gamma\delta$ T cell, 32
T cell immunity, 25–32, 32b
   through antigen-specific receptors, 25–26, 25b, 26f, 27f
   T cell development in, 27–29, 30f
   T cell phenotype functions in, 29–32
      CTLs, 29–31, 31b, 31f
      NKT cells, 32, 32b
      $\gamma\delta$ T cells, 32
      T regulatory cells, 31–32, 32b
      T_H cells, 29, 29b, 30f, 31t
   TCR genes in, 26–27, 26b, 28f, 29f

T cytotoxic cell, 10–11
  development of, 54, 56f
  functions of, 29–31, 31b, 31f
T helper (T_H) cell, 10
  functions of, 29, 29b, 30f, 31t
  phenotype development of, 53, 53b, 54f
T regulatory cells, 31–32, 32b, 62, 63f
T suppressor cells, 10–11, 31–32, 32b
Tacrolimus, 64t
Taenia, 152
TAP. *See* Transporters of antigenic peptides
Tapeworm, 152
T-cell receptors (TCRs), 25–26, 25b, 26f, 27f
  genes coding for, 26–27, 26b, 28f, 29f
  superantigen binding with, 39–40, 40f
TCRs. *See* T-cell receptors
Tetanus, 111, 112f
T_H cell. *See* T helper cell
Thymic education, 38, 38b
Thymocytes, 12–13, 38, 38b
Thymus gland, 12–13, 12f, 39b
Tick-borne encephalitis, 135
TLRs. *See* Toll-like receptors
Tolerance, immunologic, 61–63, 63b
  lymphocyte tolerance, 62, 63f
  self-reactive lymphocyte elimination, 62–63
Toll-like receptors (TLRs), 46, 48f
Tonsils, 15, 15f
Toxins
  bacterial, 100–102, 100f, 101t, 102b
  fungal, 139–141, 145–146, 146f
Toxoplasma, 149
Transcription. *See* Gene transcription
Transduction, 98, 98f
Transformation, 98, 98f
Translation. *See* RNA translation
Transplantation, immune-mediated rejection of, 65–66, 66b, 67f, 68t
Transporters of antigenic peptides (TAP), 36b, 37f
Transposons, 98
Trematodes, 152–153, 153b, 153f
Treponema, 75–76, 117
Trichinella, 85, 152
Trichomonas, 151
Trichophyton, 143
Trichosporon beigelii, 143
Trypanosoma, 150–151
Tuberculosis, 58, 58f, 82t, 83, 84f, 85b, 119, 119f
Tularemia, 115
Tumor immunity, 66, 66b
Type I hypersensitivity, 54–55, 56f
Type II hypersensitivity, 55–56, 57f
Type III hypersensitivity, 56–58, 57f
Type IV hypersensitivity, 58–59, 58f
Tyrosine kinases, 25b

U
Ulceration, 115
Urticaria, 85b

V
V genes. *See* Variable genes
Vaccination, 68–69, 69b, 69f, 69t
  antibacterial, 21b
  antiviral, 125–128, 127t
  cross-reactivity and, 21
Variable (V) genes, recombination of
  for antibodies, 21–23, 22f, 24b
  for TCRs, 26–27, 26b, 28f, 29f
Variable heavy domain (V_H), 17
Variable light domain (V_L), 17
Varicella, 131–132
Varicella-zoster virus (VZV), 131–132
Variola, 132
Vascular development, apoptosis and, 63b
Vasculitis, autoimmune, 58b
Vasoactive mediators, 9b
Veillonella, 109
Vesicular stomatitis virus (VSV), 135
V_H. *See* Variable heavy domain
Vibrio, 115, 116f
Viral infection, 121–128, 121b, 128b, 129–138, 130t, 138b
  arboviruses, 130t, 135–136
    dengue virus, 135
    Japanese encephalitis virus, 135
    tick-borne encephalitis, 135
    VSV, 135
    West Nile virus, 135
    yellow fever, 135
  childhood exanthems, 130t, 131–132
    mumps, 131
    parvovirus, 131
    rubella, 131
    rubeola, 131
    varicella, 131–132
  classification and structure of viruses, 121, 121b, 122f
  diagnostic, 123–124, 127f
  disease patterns and pathogenesis, 123, 123b, 126f
  emerging pathogens, 130t, 136–137, 137t
  enteroviruses, 130t, 132
    coxsackievirus, 132
    echoviruses, 132
    poliovirus, 132
  genetic material of viruses, 121–122, 123f
  hepatitis viruses, 130t, 132, 133f
  herpes virus, 130t, 132–134, 133t
    cytomegalovirus, 133–134, 133t
    EBV, 88–89, 133t, 134
    HHSV1 and HHSV2, 130t, 132–133, 133t, 134f
    HHSV7 and HHSV8, 133t, 134

Viral infection (*Continued*)
    HHV6, 131–132, 133t
  infectivity and replication strategies, 122–123, 123b, 124f, 125t
  major immune defense mechanisms against, 82t, 83–85, 84f
  papovaviruses, 130t, 134–135
    HPVs, 134–135, 134t
    JC and BK viruses, 135
  poxviruses, 130t, 132
  prions, 130t, 136
  respiratory viruses, 129–131, 130t
    adenovirus, 129
    coronavirus, 131
    influenza, 130
    parainfluenza, 130
    respiratory syncytial virus, 131
    rhinovirus, 131
  retroviruses, 130t, 135
    HIV, 73, 74f, 135, 136f, 137f
    HTLV, 135
  rhabdoviruses, 130t, 134
    Lyssavirus, 134
  therapy and prophylaxis for, 124–128, 126b, 127t
Virion, 121
Virulence factors, bacterial toxins, 100–102, 100f, 101t, 102b
V(D)J recombinase, 21, 22
V_L. *See* Variable light domain
VSV. *See* Vesicular stomatitis virus
VZV. *See* Varicella-zoster virus

W
Warts, 134–135
Wasserman reaction, 75–76
West Nile virus, 135
Western blot, 73, 74f
Whipworms, 151

X
Xenograft, 65

Y
Yellow fever, 135
Yersinia, 113

Z
Zoonotic-origin organisms, 114–115
  Bartonella, 115
  Brucella, 115
  Francisella, 115
  Pasteurella, 114–115
  Vibrio, 115, 116f
Zymogens, 48